Universitext

Universitext

Editors (North America): J.H. Ewing, F.W. Gehring, and P.R. Halmos

Joel L. Schiff

Normal Families

Springer-Verlag
New York Berlin Heidelberg London Paris
Tokyo Hong Kong Barcelona Budapest

Joel L. Schiff
Department of Mathematics and Statistics
University of Auckland
Private Bag 92019
Auckland, New Zealand

AMS Subject Classifications (1991): 30Dxx, 30D45, 58F08

With seven figures.

Library of Congress Cataloging-in-Publication Data
Schiff, Joel L.
    Normal families / Joel L. Schiff.
       p.   cm. — (Universitext)
    Includes bibliographical references and index.
    ISBN 0-387-97967-0
    1. Analytic functions.   2. Functions, Meromorphic.   I. Title.
  QA331.S36   1993
  515 — dc20   92-35795

Printed on acid-free paper.

Production managed by Hal Henglein; manufacturing supervised by Vincent R. Scelta.
Photocomposed copy prepared from the author's LaTeX files.
Printed and bound by Edwards Brothers, Inc., Ann Arbor, MI.
Printed in the United States of America.

9 8 7 6 5 4 3 2 1

ISBN 0-387-97967-0 Springer-Verlag New York Berlin Heidelberg
ISBN 3-540-97967-0 Springer-Verlag Berlin Heidelberg New York

Dedicated to

Paul Montel (1876–1975)

Photo reprinted with permission of Birkhäuser Boston from George Pólya, *The Pólya Picture Album: Encounters of a Mathematician*, edited by G.L. Alexanderson, 1987, p. 91.

La théorie des familles normales de fonctions doit être considérée comme l'une des découvertes les plus belles, et les plus importantes par sa fécondité, de cette première moitié du siècle.

Henri Milloux
Selecta — 1947

Preface

A book on the subject of normal families more than sixty years after the publication of Montel's treatise *Leçons sur les familles normales de fonctions analytiques et leurs applications* is certainly long overdue. But, in a sense, it is almost premature, as so much contemporary work is still being produced. To misquote Dickens, this is the best of times, this is the worst of times. The intervening years have seen developments on a broad front, many of which are taken up in this volume. A unified treatment of the classical theory is also presented, with some attempt made to preserve its classical flavour.

Since its inception early this century the notion of a normal family has played a central role in the development of complex function theory. In fact, it is a concept lying at the very heart of the subject, weaving a line of thought through Picard's theorems, Schottky's theorem, and the Riemann mapping theorem, to many modern results on meromorphic functions via the Bloch principle. It is this latter that has provided considerable impetus over the years to the study of normal families, and continues to serve as a guiding hand to future work. Basically, it asserts that a family of analytic (meromorphic) functions defined by a particular property, $\mathcal{P}$, is likely to be a normal family if an entire (meromorphic in $\mathbb{C}$) function possessing $\mathcal{P}$ reduces to a constant.

An illustration of the Bloch principle, stemming from the little Picard theorem, is the fundamental theorem of normal families, namely the *Critère fondamental* of Montel: a family of analytic (meromorphic) functions that omit two (three) particular values is normal. This theorem is presented from five differing points of view, which typify the main currents of thought within the subject. Firstly, there is the classical proof of Montel [1912] utilizing the elliptic modular function and intertwined with the ideas of Picard and Schottky; indeed, Schottky's theorem provides us with a second proof. A third proof (Chapter 3) is derived from the Ahlfors theory covering surfaces and the notion of an invariant normal family due to Hayman. Another proof, in Chapter 4, by Drasin [1969] employs the Nevanlinna theory, which has proved so indispensable in modern function theory and is the language in which much of the contemporary work on normal families is written. Finally, we present an elegantly simple proof based on the Robinson-Zalcman formalization of the Bloch principle (Robinson [1973], Zalcman [1975]).

The seeds for the notion of a normal family lie in the Bolzano-Weierstrass property: every bounded infinite set of points has a point of accumulation. With regard to families of functions, this idea was nurtured along by the work of three Italians — Ascoli, Arzelà, and Vitali — around the turn of the century. A family of continuous functions was compact if and only if it was equicontinuous and uniformly bounded. These concepts were exploited by Arzelà in an attempt to prove the Dirichlet principle, which was finally, in 1900, given a rigorous justification by Hilbert, who also used the notion of compactness. The term *compact* was introduced by Frèchet in 1904. Three years later, inspired by this circle of ideas, Montel initiated his study of a normal family of analytic functions and proved the first of many sufficient conditions he was to give over the next twenty years, that of local boundedness. Montel subsequently discovered the important connection between normal families and the celebrated theorems of Picard, Schottky, and Landau. In a series of long papers, culminating in his monumental work of 1927, he firmly established the theory of normal families and elucidated its many widespread applications.

One very significant application was made three decades later by Lehto and Virtanen [1957a], following on from the work initiated by Noshiro [1938]. This was the study of the boundary behaviour of *normal functions*, the condition of normality being exactly the requirement needed in order to extend the Lindelöf property to meromorphic functions. The importance of this class of functions had, in fact, been anticipated by Hayman [1955] in his investigation of normal invariant families.

The new notion of normal families early on became inextricably linked to the study of iterations of complex mappings, initially undertaken by Julia [1918] and Fatou [1919, 1920]: the Julia set is just the set of points at which the family of iterations is *not* normal, and the Fatou set is its complement. One of the ironies of the subject of normal families is that the study of Julia sets, having lain dormant for several decades, has once again attracted renewed interest because of their importance in the study of dynamical systems. *Plus ça change, plus c'est la même chose!* Moreover, although it was known from Fatou and Julia's early work that for rational functions the Julia set coincided with the closure of the set of repelling fixed points, it was only in 1968 that Baker established the corresponding result for entire functions. His proof, presented in Chapter 5, embraces once again the Ahlfors theory of covering surfaces and provides a further link between this beautiful theory and that of normal families.

Naturally, in a book of this type certain compromises have to be made. Various theorems are sometimes merely stated, often because their proofs are of considerable length and engender the use of numerous highly technical lemmas. This is particularly true of theorems involving the Nevanlinna theory, where lengthy inequalities are frequently encountered, with nearly every term requiring a further delicate estimate. Therefore, a middle course has been attempted by giving proofs of a substantial core of work, with indications as to generalizations and related results.

For reasons of length, the subject matter has been restricted to the study of normal families of analytic and meromorphic functions. Topics omitted include normal family complexes, normal families of quasi-conformal mappings, quasi-regular mappings, and functions of real variables and of several complex variables, amongst others. These await a sequel to the present volume. However, the book does contain some of the most beautiful mathematics of this century. I hope that I have done it justice.

Of course, I owe a great debt of gratitude to the many individuals who have assisted me with this work over the years. I would first like to mention that it was Tony Holland, who shortly before he died gave me the idea to write this book. Parts of early versions of the manuscript were read by graduate students Matthew Bell, Simon Marchant, Geoff Pritchard, and Paul Taylor. Parts of the final version were read by Professors Jim Clunie, David Drasin, Walter Hayman, Norman Levenson, Yang Lo, Gaven Martin, and Lawrence Zalcman. All of them made valuable suggestions and contributions which considerably improved the text. In particular, the unstinting assistance of Professor Walter Hayman at every stage made me realise that he is not only a brilliant mathematician but a great human being as well. I am also grateful for the numerous obscure references from the British Library that were furnished by Dr. Nick Dudley Ward. Part of this work was written while on sabbatical leave at the University of York during 1989, where I was taken in and treated like one of the staff. The typing has been patiently and expertly done by Lois Kennedy and Betty Fong. Thanks are also due to Dr. Colin Fox, who produced Figures 5.1 and 5.2, and to Bernadette Wehr of Springer-Verlag for her generous support.

Finally, I would like to thank my wife, Christine, and daughter, Averil, for their support and understanding. To my son, Aaron, I give a special thanks for his expert technical assistance. Perhaps now we can return to being a normal family again.

Contents

1

Preliminaries

In this chapter we present notational conventions and some standard results which commonly feature in the text. Some elementary knowledge of complex variable theory and topology is assumed.

1.1 Basic Notation

$\mathbb{Z}$: integers.

$\mathbb{N}$: natural numbers $1, 2, 3 \ldots$

$\mathbb{R}$: real numbers.

$\mathbb{C}$: complex plane $|z| < \infty$.

$\widehat{\mathbb{C}}$: extended complex plane, $\mathbb{C} \cup \{\infty\}$, identified with the Riemann sphere, Σ.

$\mathcal{H}_+$: upper half-plane.

$\mathcal{H}_-$: lower half-plane.

U : open unit disk $|z| < 1$.

$D(z_0; r)$: an open disk about $z_0 \in \mathbb{C}$; $\{z : |z - z_0| < r\}$, $r > 0$.

$D'(z_0; r)$: a deleted (punctured) disk about z_0; $D(z_0; r) - \{z_0\}$.

$K(z_0; r)$: a closed disk about z_0; $\{z : |z - z_0| \le r\}$, $r > 0$.

Ω : a domain (connected open set) in $\mathbb{C}$. In some specified instances, the domain will be located on the Riemann sphere.

$\{f_n\}$: a sequence of functions, i.e., an ordered countably infinite collection of distinct functions. Throughout the text, the range of an otherwise unspecified sequence $\{f_n\}$ is taken lie in $\mathbb{C}$ or on Σ, and the context will make it clear which one is appropriate.

$A, B, C, \ldots$: constants. The same letter is frequently used for different constant values, when no confusion arises.

$\overline{E}, E^0, \partial E, E', \overline{\overline{E}}$: the closure, interior, boundary, complement, cardinality, of a set E.

Additional terms are defined throughout the text. Good general references for much of the elementary material to follow can be found in Ahlfors [1979], Behnke and Sommer [1965], Hille [1962, Vols. I, II], and Palka [1991].

1.2 Spherical and Hyperbolic Metrics

The Spherical Metric. Consider in $\mathbf{R}^3$ the sphere Σ : $x_1^2 + x_2^2 + \left(x_3 - \frac{1}{2}\right)^2 = \frac{1}{4}$. Then the $x_1 x_2$-plane is tangent to Σ at $(0, 0, 0)$, and $N = (0, 0, 1)$ is the *north pole*. A line from N to a point z_1 in the (complex) $x_1 x_2$-plane intersects the sphere at a point $P_1\,(\alpha_1, \beta_1, \gamma_1)$. This sets up a one-to-one correspondence $z_1 \leftrightarrow P_1$ between the points of $\mathbb{C}$ and $\Sigma - \{N\}$. Associating N with ∞, we extend the correspondence to one between Σ and $\widehat{\mathbb{C}} = \mathbb{C} \cup \{\infty\}$ (cf. Figure 1.1). We call Σ the *Riemann sphere*, $\widehat{\mathbb{C}}$ the *extended complex plane*, and the correspondence a *stereographic projection*. This mapping is conformal (i.e., angle-preserving), and so the corresponding conformal geometries are the same. The identification of $\widehat{\mathbb{C}}$ with Σ allows us to use them interchangeably.

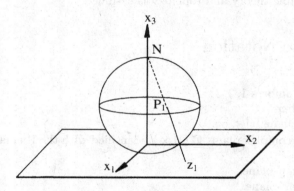

Figure 1.1

Let z_1 and z_2 be the two points in $\mathbb{C}$ corresponding to P_1 and P_2, respectively, on Σ. If $P_i = (\alpha_i, \beta_i, \gamma_i)$, $i = 1, 2$, then the Euclidean distance between P_1 and P_2 is given by

$$|P_1 - P_2| = \left[(\alpha_1 - \alpha_2)^2 + (\beta_1 - \beta_2)^2 + (\gamma_1 - \gamma_2)^2\right]^{\frac{1}{2}}.$$

We denote $|P_1 - P_2|$ by $\chi(z_1, z_2)$, the *chordal* distance between z_1 and z_2. It can be shown that if z_1 and z_2 are in the finite plane, then

$$\chi(z_1, z_2) = \frac{|z_1 - z_2|}{\sqrt{1 + |z_1|^2}\sqrt{1 + |z_2|^2}},$$

and if $z_2 = \infty$, then

$$\chi(z_1, \infty) = \frac{1}{\sqrt{1 + |z_1|^2}}.$$

Clearly, $\chi(z_1, z_2) \leq 1$,

$$\chi\left(\frac{1}{z_1}, \frac{1}{z_2}\right) = \chi(z_1, z_2),$$

and if $|z_1| \leq |z_2| \leq \infty$, then $\chi(0, z_1) \leq \chi(0, z_2)$. It is also true that $\chi(\cdot, \cdot)$ is a *metric* on $\widehat{\mathbb{C}}$, with $\chi(z_1, z_2) \leq |z_1 - z_2|$ in $\mathbb{C}$. The virtue of the chordal metric is that it allows $z = \infty$ to be treated like any other point.

The *spherical arc length element ds* on the Riemann sphere Σ works out to be

$$ds = \frac{|dz|}{1 + |z|^2},$$

and the *spherical area element dA* is given by

$$dA = \frac{dx\,dy}{(1 + |z|^2)^2} \qquad (z = x + i\,y).$$

The spherical length

$$L(\gamma) = \int_\gamma \frac{|dz|}{1 + |z|^2}$$

of a curve γ on Σ induces a metric in the following manner. Given distinct points z_1, z_2 on the Riemann sphere, define

$$\sigma(z_1, z_2) = \inf\{L(\gamma)\},$$

where the infimum is taken over all differentiable curves on Σ which join z_1 with z_2. Then $\sigma(z_1, z_2)$ is the euclidean length of the shortest arc of the great circle on Σ joining z_1 and z_2 and defines a metric on the sphere known as the *spherical metric*. Indeed, $\chi(z_1, z_2) \leq \sigma(z_1, z_2) \leq \frac{\pi}{2}\chi(z_1, z_2)$, so that the two metrics are *uniformly equivalent* and generate the same open sets on Σ. Thus, from a topological point of view, the metrics χ and σ can be treated as one and the same.

Definition 1.2.1 *A sequence of functions $\{f_n\}$ converges **spherically uniformly** to f on a set $E \subseteq \mathbb{C}$ if, for any $\varepsilon > 0$, there is a number n_0 such that $n \geq n_0$ implies*

$$\chi\big(f(z), f_n(z)\big) < \varepsilon,$$

for all $z \in E$.

Note that if $\{f_n\}$ converges uniformly to f on E, then it also converges spherically uniformly to f on E. The converse holds if the limit function is bounded.

Theorem 1.2.2 *If the sequence $\{f_n\}$ converges spherically uniformly to a bounded function f on E, then $\{f_n\}$ converges uniformly to f on E.*

Proof. Suppose $|f(z)| \le M$ on E. Then

$$\chi(0, f(z)) \le \chi(0, M) = \frac{M}{\sqrt{1 + M^2}} < 1.$$

Choose $\varepsilon < 1 - \frac{M}{\sqrt{1+M^2}}$. Then there exists a positive integer n_0 such that $n \ge n_0$ implies

$$\chi(f(z), f_n(z)) < \varepsilon,$$

so that

$$\frac{|f_n(z)|}{\sqrt{1 + |f_n(z)|^2}} = \chi(0, f_n(z)) \le \chi(0, f(z)) + \chi(f(z), f_n(z))$$

$$< \frac{M}{\sqrt{1 + M^2}} + \varepsilon = m < 1.$$

We deduce that

$$|f_n(z)| < \frac{m}{\sqrt{1 - m^2}} = M_1,$$

for all $n \ge n_0$. Hence

$$|f(z) - f_n(z)| = \sqrt{1 + |f(z)|^2}\sqrt{1 + |f_n(z)|^2} \cdot \chi(f(z), f_n(z))$$
$$< \sqrt{1 + M^2}\sqrt{1 + M_1^2} \cdot \chi(f(z), f_n(z)),$$

$n \ge n_0$. The uniform convergence then follows from this latter inequality.

As to the notion of continuity with respect to the chordal metric, we have

Definition 1.2.3 *A function f is **spherically continuous** at a point $z_0 \in \mathbb{C}$ if, given $\varepsilon > 0$, there exists $\delta > 0$ such that*

$$\chi(f(z), f(z_0)) < \varepsilon,$$

whenever $|z - z_0| < \delta$.

In the case of meromorphic functions this leads to:

Proposition 1.2.4 *If $f(z)$ is meromorphic in a domain Ω, then f is spherically continuous in Ω.*

Proof. If $f(z)$ is analytic at $z_0 \in \Omega$, then it is spherically continuous there since

$$\chi(f(z), f(z_0)) < |f(z) - f(z_0)|.$$

If z_0 is a pole, then $\frac{1}{f(z)}$ is continuous at z_0, and noting that

$$\chi(f(z), f(z_0)) = \chi\left(\frac{1}{f(z)}, \frac{1}{f(z_0)}\right),$$

the result follows as in the preceding case.

Let $f(z)$ be meromorphic on a domain Ω. If $z \in \Omega$ is not a pole, the derivative in the spherical metric, called the *spherical derivative*, is given by

$$
\begin{aligned}
f^{\#}(z) &= \lim_{z' \to z} \frac{\chi(f(z), f(z'))}{|z - z'|} \\
&= \lim_{z' \to z} \frac{|f(z) - f(z')|}{|z - z'|} \cdot \frac{1}{\sqrt{1 + |f(z)|^2}} \cdot \frac{1}{\sqrt{1 + |f(z')|^2}} \\
&= \frac{|f'(z)|}{1 + |f(z)|^2}.
\end{aligned}
$$

If ζ is a pole of $f(z)$, define

$$
f^{\#}(\zeta) = \lim_{z \to \zeta} \frac{|f'(z)|}{1 + |f(z)|^2}.
$$

Thus $f^{\#}(z)$ is continuous, and one can verify that $f^{\#}(z) = \left(\frac{1}{f(z)}\right)^{\#}$. If γ is a differentiable arc or curve in Ω, then the image of γ on the Riemann sphere has arc length element $ds = f^{\#}(z)|dz|$, and its spherical length is given by

$$
\int_{\gamma} f^{\#}(z)|dz|.
$$

In particular, if $f(z)$ is meromorphic in $|z| \leq r$, denote

$$
L(r) = \int_{|z|=r} f^{\#}(z)|dz| = \int_0^{2\pi} \frac{|f'(re^{i\theta})|r\, d\theta}{1 + |f(re^{i\theta})|^2}, \tag{1.1}
$$

$$
\begin{aligned}
S(r) &= \frac{1}{\pi} \int_{|z|<r} \int \frac{|f'(z)|^2\, dx\, dy}{(1 + |f(z)|^2)^2} \\
&= \frac{1}{\pi} \int_0^{2\pi} \int_0^r \frac{|f'(\rho e^{i\theta})|^2 \rho\, d\rho\, d\theta}{\left(1 + |f(\rho e^{i\theta})|^2\right)^2}, \tag{1.2}
\end{aligned}
$$

i.e., $L(r) =$ the length of the image of the circle $\{z : |z| = r\}$ on the Riemann sphere; $S(r) = \frac{1}{\pi} \cdot$ area of the image of the disk $\{z : |z| < r\}$ on the Riemann sphere, determined with regard for multiplicity.

The Hyperbolic Metric. We turn our attention now to the open unit disk U. Define the *hyperbolic arc length* element by

$$
d\varsigma = \frac{|dz|}{1 - |z|^2},
$$

in contrast with the arc length element $ds = |dz|/(1 + |z|^2)$ we encountered with the spherical metric. If γ, parametrized by $z(t) : [a, b] \to U$, is a differentiable arc or curve in U, then for $z = z(t)$, $|dz| = |z'(t)| dt$, the hyperbolic length of γ is given by

$$\lambda(\gamma) = \int_\gamma d\varsigma = \int_a^b \frac{|z'(t)| \, dt}{1 - |z(t)|^2}.$$

Next, let $w = \phi(z)$ be a one-to-one conformal mapping of U onto itself. By virtue of the Schwarz-Pick Lemma (cf., e.g., Goluzin [1969], p. 332),

$$\int_{\phi(\gamma)} \frac{|dw|}{1 - |w|^2} \leq \int_\gamma \frac{|dz|}{1 - |z|^2},$$

that is, $\lambda(\phi(\gamma)) \leq \lambda(\gamma)$. A second application of Schwarz-Pick to ϕ^{-1} gives $\lambda(\gamma) \leq \lambda(\phi(\gamma))$, which shows that ϕ is length-preserving.

Let us compute the hyperbolic length of the straight-line segment ℓ connecting 0 with r, $0 < r < 1$. Indeed

$$\lambda(\ell) = \int_0^r \frac{dt}{1 - t^2} = \frac{1}{2} \log \frac{1 + r}{1 - r}.$$

This is significant as ℓ is a *geodesic*, i.e., a curve of minimum hyperbolic length between 0 and r. To verify this, let γ be an arbitrary differentiable arc joining 0 and r, parametrized by $z(t) = x(t) + iy(t)$, $0 \leq t \leq r$. Then

$$\lambda(\gamma) = \int_0^r \frac{|z'(t)| \, dt}{1 - |z(t)|^2} \geq \int_0^r \frac{x'(t) \, dt}{1 - [x(t)]^2} = \frac{1}{2} \log \frac{1 + r}{1 - r} = \lambda(\ell),$$

whence the assertion. Note that the inequality is strict unless $y(t) \equiv 0$, so that the geodesic ℓ is unique.

This fact can then be used to determine the geodesic joining any two points a, $b \in U$. Let F be a Möbius transformation mapping $\overline{U}$ onto $\overline{U}$ such that $F(a) = 0$, $F(b) = r$. The geodesic connecting a and b is the inverse image of the line segment ℓ connecting 0 and r, since F is length-preserving. This inverse transformation takes circles into circles, regarding a straight line as a special case of a circle. Furthermore, since the mapping is conformal and the line segment ℓ connecting 0 and r is orthogonal to $|z| = 1$, so is the geodesic joining a and b. Hence the geodesic between a and b is a circular arc joining them, which, if extended, meets the circle $|z| = 1$ orthogonally.

The preceding considerations provide a model for noneuclidean geometry, but this aspect will not be pursued here (cf. Beardon [1983], Chapter 7).

Define

$$\phi(z) = \frac{z - z_1}{1 - \overline{z}_1 z}.$$

Then ϕ is a Möbius transformation which satisfies $\phi(z_1) = 0$, with $|\phi(z_2)|$ $= r$, $0 < r < 1$, for fixed z_1, $z_2 \in U$. If γ is the geodesic joining z_1 and z_2, then

$$\lambda(\gamma) = \lambda\big(\phi(\gamma)\big) = \frac{1}{2} \log \frac{1+r}{1-r}.$$

But

$$r = |\phi(z_2)| = \left| \frac{z_1 - z_2}{1 - \overline{z}_1 z_2} \right|,$$

which leads to

$$\begin{aligned} \lambda(\gamma) &= \log \frac{1 + \left| \frac{z_1 - z_2}{1 - \overline{z}_1 z_2} \right|}{1 - \left| \frac{z_1 - z_2}{1 - \overline{z}_1 z_2} \right|} \\ &\equiv \rho(z_1, z_2). \end{aligned}$$

It is evident that $\rho(z_1, z_2)$, being the length of the geodesic joining z_1 and z_2, is a metric, called the *hyperbolic metric*.

A *hyperbolic disk*, with *hyperbolic centre* z_0 and *hyperbolic radius* r is defined by

$$\mathcal{D}(z_0; r) = \{ z \in U : \rho(z_0, z) \le r \}.$$

Then $z \in \mathcal{D}(z_0; r)$ entails

$$\left| \frac{z - z_0}{1 - \overline{z}_0 z} \right| \le \frac{e^{2r} - 1}{e^{2r} + 1} = \tanh r < 1,$$

which describes a euclidean disk in U whose centre differs from z_0.

1.3 Normal Convergence

The prime mode of convergence in the theory of normal families is given by

Definition 1.3.1 *A sequence of functions* $\{f_n\}$ **converges (spherically) uniformly on compact subsets** *of a domain* Ω *to a function* $f(z)$ *if, for any compact subset* $K \subseteq \Omega$ *and* $\varepsilon > 0$, *there exists a number* $n_0 = n_0(K, \varepsilon)$ *such that* $n \ge n_0$ *implies*

$$|f_n(z) - f(z)| < \varepsilon, \qquad \big(\chi\big(f_n(z), f(z)\big) < \varepsilon \big),$$

for all $z \in K$.

When a sequence converges uniformly or spherically uniformly on compact subsets of a domain Ω, we say the sequence converges *normally* in Ω when no confusion arises.

8   1. Preliminaries

On the vector space $A(\Omega)$ of all analytic functions on the domain Ω, there is a metric defined by

$$d(f,g) = \sum_{\nu=1}^{\infty} \frac{1}{2^{\nu}} \inf\big(1, \sup_{z \in K_{\nu}} |f(z) - g(z)|\big), \qquad f, g \in A(\Omega),$$

(cf., e.g., Cartan [1963], p. 146). Here $\{K_{\nu}\}$ is an *exhaustion* by compact subsets of Ω, i.e., $K_{\nu} \subseteq \Omega$, $K_{\nu} \subseteq K_{\nu+1}$, and for any compact subset $K \subseteq \Omega$, $K \subseteq K_{\nu}$ for some ν. Thus we have

Theorem 1.3.2 *Uniform convergence on compact subsets of a domain Ω is equivalent to convergence in the metric d.*

Proof. Suppose that $d(f_n, f) \to 0$ as $n \to \infty$. Let $K \subseteq \Omega$ be compact and K_{μ} a member of the exhaustion such that $K_{\mu} \supseteq K$. For $0 < \varepsilon < 1$, choose $\delta > 0$ satisfying $\delta \cdot 2^{\mu} < \epsilon$.

Then for some n_0,

$$\sum_{\nu=1}^{\infty} \frac{1}{2^{\nu}} \inf\big(1, \sup_{z \in K_{\nu}} |f_n(z) - f(z)|\big) < \delta$$

for $n \geq n_0$. Hence

$$\inf\big(1, \sup_{z \in K_{\mu}} |f_n(z) - f(z)|\big) < \varepsilon, \quad n \geq n_0,$$

and we conclude that $\sup_{z \in K} |f_n(z) - f(z)| < \varepsilon$, as desired. On the other hand, suppose that $f_n \to f$ uniformly on compact subsets of Ω, and let $\{K_{\nu}\}$ be an exhaustion of Ω. Since for any $n \in \mathbb{N}$, the terms of

$$\sum_{\nu=1}^{\infty} \frac{1}{2^{\nu}} \inf\big(1, \sup_{z \in K_{\nu}} |f_n(z) - f(z)|\big)$$

are dominated by $\frac{1}{2^{\nu}}$, the Weierstrass M-Test implies there exists $\nu_0 \in \mathbb{N}$ (independent of n) such that

$$\sum_{\nu=\nu_0}^{\infty} \frac{1}{2^{\nu}} \inf\big(1, \sup_{z \in K_{\nu}} |f_n(z) - f(z)|\big) < \frac{\epsilon}{2}.$$

Moreover, there exists $n_0 \in \mathbb{N}$ such that $n \geq n_0$ implies

$$\sum_{\nu=1}^{\nu_0-1} \frac{1}{2^{\nu}} \big(\sup_{z \in K_{\nu}} |f_n(z) - f(z)| \big) < \frac{\epsilon}{2}.$$

Therefore, $n \geq n_0$ gives $d(f_n, f) < \epsilon$ and the proof is complete.

It follows that the topology on $A(\Omega)$ induced by uniform convergence on compact subsets is defined by the metric d. If $f_n \to f$ uniformly on each closed disk $K \subseteq \Omega$, then we say that $\{f_n\}$ converges *locally uniformly* to f in Ω. Using the Heine-Borel Theorem, local uniform convergence is easily seen to be equivalent to uniform convergence on compact subsets.

1.4 Some Classical Theorems

The next four classical results of Weierstrass, Hurwitz, and Schwarz find frequent application throughout the text.

Weierstrass Theorem *Let $\{f_n\}$ be a sequence of analytic functions on a domain Ω which converges uniformly on compact subsets of Ω to a function f. Then f is analytic in Ω, and the sequence of derivatives $\{f_n^{(k)}\}$ converges uniformly on compact subsets to $f^{(k)}$, $k = 1, 2, 3 \ldots$.*

Proof. For an arbitrary $z_0 \in \Omega$, choose a disk $D(z_0; r) \subseteq \Omega$, and write $C_r = \{z : |z - z_0| = r\}$. For any $\epsilon > 0$, the hypothesis implies there exists $n_0 \in \mathbb{N}$ such that if $n \geq n_0$,

$$|f_n(\zeta) - f(\zeta)| < \epsilon$$

for all $\zeta \in C_r$. Next, define

$$F_k(z) = \frac{k!}{2\pi i} \int_{C_r} \frac{f(\zeta)\, d\zeta}{(\zeta - z)^{k+1}}, \qquad k = 0, 1, 2, \ldots,$$

for $z \in D(z_0; \frac{r}{2})$. Then for $n \geq n_0$,

$$
\begin{aligned}
|F_k(z) - f_n^{(k)}(z)| &\leq \frac{k!}{2\pi} \int_{C_r} \frac{|f(\zeta) - f_n(\zeta)|}{|\zeta - z|^{k+1}} |d\zeta| \\
&< \frac{k!\epsilon 2^{k+1}}{r^k},
\end{aligned}
$$

that is, $f_n^{(k)}(z) \to F_k(z)$ uniformly on $D\left(z_0; \frac{r}{2}\right)$, $k = 0, 1, 2, \ldots$. For $k = 0$, $f_n(z) \to f(z)$ and $f_n(z) \to F_0(z)$ give $f(z) \equiv F_0(z)$, which is an analytic function in $D\left(z_0; \frac{r}{2}\right)$. As z_0 was arbitrary, $f(z)$ is analytic in Ω. Then $f_n^{(k)} \to f^{(k)}$ uniformly on $D\left(z_0; \frac{r}{2}\right)$, and the obvious compactness argument establishes the theorem.

Hurwitz Theorem *Let $\{f_n\}$ be a sequence of analytic functions on a domain Ω which converges uniformly on compact subsets of Ω to a non-constant analytic function $f(z)$. If $f(z_0) = 0$ for some $z_0 \in \Omega$, then for each $r > 0$ sufficiently small, there exists an $N = N(r)$, such that for all $n > N$, $f_n(z)$ has the same number of zeros in $D(z_0; r)$ as does $f(z)$. (The zeros are counted according to multiplicity).*

Proof. Take a disk $D(z_0; r)$ to be so small that $f(z)$ does not vanish in $K(z_0; r)$ except at z_0. Then for some $m > 0$, $|f(z)| > m$ on $|z - z_0| = r$ by the continuity of f. As $f_n \to f$ uniformly on $|z - z_0| = r$, we have

$$|f_n(z) - f(z)| < m < |f(z)|, \qquad |z - z_0| = r,$$

for all n sufficiently large. The conclusion follows by virtue of Rouché's theorem.

A complex-valued function which is one-to-one, that is, if $z_1 \neq z_2$, then $f(z_1) \neq f(z_2)$, is termed *univalent*.

Corollary 1.4.1 *If* $\{f_n\}$ *is a sequence of univalent analytic functions in a domain* Ω *which converge uniformly on compact subsets of* Ω *to a nonconstant analytic function* f, *then* f *is univalent in* Ω.

Proof. Suppose that $f(z_1) = f(z_2)$ for some $z_2 \in \Omega$. Consider the functions

$$g_n(z) = f_n(z) - f_n(z_1), \quad n = 1, 2, 3, \ldots .$$

The limit function $g(z) = f(z) - f(z_1)$ vanishes at z_2 so that by the theorem, $g_n(z)$ has a zero in a neighbourhood of z_2, for all n sufficiently large. However, as each f_n is univalent, $g_n(z)$ only vanishes at z_1, and this contradiction proves the corollary.

Schwarz Lemma *Let* $f(z)$ *be analytic in* U *with* $f(0) = 0$, $|f(z)| \leq 1$. *Then* $|f(z)| \leq |z|$ *for all* $z \in U$, *and* $|f'(0)| \leq 1$. *If* $|f(z_0)| = |z_0|$ *for some* $z_0 \in U$, *then* $f(z) \equiv cz$, *where* c *is a constant of modulus 1.*

Proof. Because $f(0) = 0$, we can write $f(z) = zg(z)$, for some $g(z)$ analytic in U. On the circle $|z| = r$, $0 < r < 1$, we have

$$|g(z)| = \left| \frac{f(z)}{z} \right| \leq \frac{1}{r},$$

so that by the Maximum Modulus Principle, $|g(z)| \leq \frac{1}{r}$ in $|z| \leq r$. By letting r tend to 1, we obtain $|g(z)| \leq 1$, for all $z \in U$, as desired. Moreover, $|f'(0)| = |g(0)| \leq 1$. Finally, if $|g(z_0)| = 1$ for some $z_0 \in U$, then again by the Maximum Principle, $g(z) \equiv c$, where $|c| = 1$.

The general form of the Schwarz inequality for $f(z)$ analytic in $|z| < R$ and $|f(z)| \leq M$ is readily verified to be

$$|f(z)| \leq \frac{M}{R}|z|.$$

At the very core of Picard's theorems, the theorems of Landau and Schottky, as well as Montel's Fundamental Normality Test (amongst others) is an essential result due to Weierstrass known as the Monodromy Theorem.

Given two overlapping open disks Δ_1 and Δ_2, suppose that there are functions $f_1(z)$, $f_2(z)$ which are analytic in Δ_1, Δ_2, respectively, and that $f_1(z) = f_2(z)$ for all $z \in \Delta_1 \cap \Delta_2$. The pairs $(f_i(z), \Delta_i)$ are called *function elements*, and $f_2(z)$ is said to be a *direct analytic continuation of* $f_1(z)$

into Δ_2 and vice versa, thereby defining an analytic function in $\Delta_1 \cup \Delta_2$. Extending this idea, suppose AB is a closed line segment contained in the union of a chain of finitely many such disks, Δ_i, $i = 1, \ldots, n$. Further, assume there are analytic functions $f_i(z)$ defined on each Δ_i, so that $f_{i+1}(z)$ is the direct continuation of $f_i(z)$ to Δ_{i+1}. Then any such pair of function elements in the chain are *analytic continuations* of each other, engendering a single-valued analytic function $f(z)$ being defined on $\cup_{i=1}^n \Delta_i$. A similar procedure can be adopted in the case of an arbitrary polygonal path, a fundamental consequence of which is the

Monodromy Theorem *Let Ω be a simply connected domain and fix a point $z_0 \in \Omega$. Suppose that $f_0(z)$ is an analytic function defined in a neighbourhood of z_0, and that $f_0(z)$ can be analytically continued along any polygonal path issuing from z_0 and lying entirely in Ω. Then the aggregate of all such continuations defines a single-valued analytic function in the whole of Ω.*

Proof. Note first that the domain Ω is *polygonally connected*, that is, any two points z_0 and z_1 in Ω can be connected by a polygonal path lying wholly in Ω with end points z_0, z_1.

Now, consider a triangle ABC lying entirely in Ω, and suppose that $f_0(z)$ is analytic in a neighbourhood of A. Then by the foregoing discussion, $f_0(z)$ can be continued analytically along any line segment AP, for P a point on BC, thus defining a single-valued analytic function in a neighbourhood of the segment AP. Within this neighbourhood, determine a triangular path AP_1PP_2A, with P_1P_2 lying on BC and P interior to the segment P_1P_2 (cf. Figure 1.2). Analytic continuation along this particular triangular path then leads to a function which coincides with $f_0(z)$ in a neighbourhood of A. By the compactness of the segment BC, there is a covering of it by a finite number of segments of the form P_1P_2. It is then deduced that $f_0(z)$ can

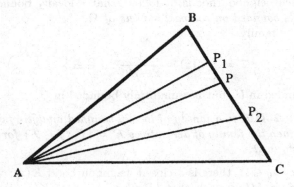

Figure 1.2

be analytically continued along the triangular path $ABCA$, with the result that the final function coincides with $f_0(z)$ at A.

We extend the preceding argument by induction on the number of sides of the polygon considered. Suppose that we have the desired analytic continuation along all closed n-gons lying in Ω. Given an arbitrary closed $(n+1)$-gon, by constructing an appropriate diagonal spanning the interior of the figure (both of which lie in Ω by the simple connectivity), it is thereby reduced to two polygons having at most n sides, and the induction hypothesis is then applicable.

As a consequence, analytic continuation of $f_0(z)$ along any two polygonal paths joining z_0 to any z_1 in Ω results in the same final function at z_1, and the theorem is proved.

Connecting z_0, z_1 in Ω by two *Jordan arcs* γ_1, γ_2 in Ω, and considering polygonal paths approximating them sufficiently closely, shows that the analytic continuation along γ_1 and γ_2 will result in the same function at z_1 as above.

1.5   Local Boundedness

There is a generalization of the concept of boundedness of a single function to that of families of functions which plays a key role in our considerations of normality.

Definition 1.5.1 *A family of functions $\mathcal{F}$ is **locally bounded** on a domain Ω if, for each $z_0 \in \Omega$, there is a positive number $M = M(z_0)$ and a neighbourhood $D(z_0; r) \subseteq \Omega$ such that $|f(z)| \leq M$ for all $z \in D(z_0; r)$ and all $f \in \mathcal{F}$.*

That is, $\mathcal{F}$ is uniformly bounded in a neighbourhood of each point of Ω. In the literature this property is sometimes referred to as *local uniform boundedness*. Since any compact subset $K \subseteq \Omega$ can be covered by a finite number of such neighbourhoods, it follows that a locally bounded family $\mathcal{F}$ is *uniformly bounded on compact subsets of Ω*.

However, the family

$$\mathcal{F} = \left\{ f_\alpha(z) = \frac{1}{z - e^{i\alpha}} : \alpha \in \mathbf{R} \right\}$$

is locally bounded in U, but not uniformly bounded in U.

Theorem 1.5.2 *If $\mathcal{F}$ is a family of locally bounded analytic functions on a domain Ω, then the family of derivatives $\mathcal{F}' = \{f' : f \in \mathcal{F}\}$ form a locally bounded family in Ω.*

Proof. For any $z_0 \in \Omega$, there is a closed neighbourhood $K(z_0; r) \subseteq \Omega$ and a constant $M = M(z_0)$ such that $|f(z)| \leq M$, $z \in K(z_0; r)$. Then for

$z \in D\left(z_0; \frac{r}{2}\right)$, and $\zeta \in C_r = \{z : |z - z_0| = r\}$, the Cauchy Formula gives

$$|f'(z)| \le \frac{1}{2\pi} \int_{C_r} \frac{|f(\zeta)||d\zeta|}{|\zeta - z|^2} < \frac{4M}{r}$$

for all $f' \in \mathcal{F}'$, so that $\mathcal{F}'$ is locally bounded.

The converse of Theorem 1.5.2 is false, as is illustrated by the family of constants $\mathcal{F} = \{n : n = 1, 2, 3, \ldots\}$.

However, the following partial converse does hold.

Theorem 1.5.3 *Let $\mathcal{F}$ be a family of analytic functions on Ω such that the family of derivatives $\mathcal{F}'$ is locally bounded and suppose that there is some $z_0 \in \Omega$ with $|f(z_0)| \le M < \infty$ for all $f \in \mathcal{F}$. Then $\mathcal{F}$ is locally bounded.*

Proof. Taking $z \in \Omega$ and integrating over a path $C \subseteq \Omega$ from z_0 to z yields

$$|f(z)| \le |f(z_0)| + \int_C |f'(\zeta)||d\zeta| \le M + M_1 \cdot \ell(C), \quad f \in \mathcal{F}.$$

The result then follows by considering an analogous inequality in a disk about the point z and integrating over any straight-line path in the disk.

1.6 Equicontinuity

Another principal notion in the development of normal families, and related to that of local boundedness, is

Definition 1.6.1 *A family $\mathcal{F}$ of functions defined on a domain Ω is said to be **equicontinuous** (**spherically equicontinuous**) at a point $z' \in \Omega$ if, for each $\epsilon > 0$, there is a $\delta = \delta(\epsilon, z') > 0$ such that*

$$|f(z) - f(z')| < \epsilon \qquad (\chi(f(z), f(z')) < \epsilon)$$

*whenever $|z - z'| < \delta$, for every $f \in \mathcal{F}$. Moreover, $\mathcal{F}$ is **equicontinuous** (**spherically equicontinuous**) on a subset $E \subseteq \Omega$ if it is equicontinuous (spherically equicontinuous) at each point of E.*

Remark. Note that if $E \subseteq \Omega$ is compact, then a simple compactness argument (based on the fact that a continuous function on a compact set is uniformly continuous) shows that if $\mathcal{F}$ is equicontinuous (spherically equicontinuous) on E, then for each $\varepsilon > 0$, there is a $\delta = \delta(\varepsilon, E) > 0$ such that

$$|f(z) - f(z')| < \varepsilon \qquad (\chi(f(z), f(z')) < \varepsilon)$$

whenever $z, z' \in E$, $|z - z'| < \delta$, for every $f \in \mathcal{F}$.

This formulation is sometimes taken as the definition of equicontinuity on an arbitrary set $E \subseteq \Omega$ (cf. Ahlfors [1979], p. 211). For compact sets then, the two approaches are equivalent. When considering spherical equicontinuity with $\Omega \subseteq \widehat{\mathbb{C}}$, we replace $|z - z'|$ by $\chi(z, z')$.

The concept of equicontinuity was originally formulated by Ascoli [1883]. Since

$$\chi\big(f(z), f(z')\big) < |f(z) - f(z')|,$$

we see that equicontinuity implies spherical equicontinuity. Furthermore, we have (cf. Theorem 2.2.1)

Proposition 1.6.2 *If $\{f_n\}$ is a sequence of spherically continuous functions which converges spherically uniformly to a function f on a compact subset $E \subseteq \mathbb{C}$, then f is uniformly spherically continuous on E, and the functions $\{f_n\}$ are spherically equicontinuous on E.*

Proof. Given $\varepsilon > 0$

$$\chi\big(f_n(z), f(z)\big) < \frac{\varepsilon}{3}, \qquad z \in E,$$

for $n \geq n_0$, for some $n_0 \in \mathbb{N}$. By the (uniform) spherical continuity of f_n on E, in particular f_{n_0}, there exists $\delta = \delta(\varepsilon, E) > 0$ such that

$$\chi\big(f_{n_0}(z), f_{n_0}(z')\big) < \frac{\varepsilon}{3},$$

whenever $z, z' \in E$, $|z - z'| < \delta$. Then

$$\begin{aligned}
\chi\big(f(z), f(z')\big) &\leq \chi\big(f(z), f_{n_0}(z)\big) + \chi\big(f_{n_0}(z), f_{n_0}(z')\big) + \chi\big(f_{n_0}(z'), f(z')\big) \\
&< \frac{\varepsilon}{3} + \frac{\varepsilon}{3} + \frac{\varepsilon}{3} = \varepsilon,
\end{aligned}$$

for $|z - z'| < \delta$, and f is uniformly spherically continuous on E. The spherical equicontinuity of $\{f_n\}$ results from an application of the triangle inequality to $\chi\big(f_n(z), f_n(z')\big)$, for $n \geq n_0$, $|z - z'| < \delta$, and the spherical continuity of f. Note that the term "spherical(ly)" may be deleted throughout.

In the present context, the relationship between local boundedness and equicontinuity is the following.

Theorem 1.6.3 *A locally bounded family $\mathcal{F}$ of analytic functions on a domain Ω is equicontinuous on compact subsets of Ω.*

Proof. By Theorem 1.5.2, $\mathcal{F}$ locally bounded implies $\mathcal{F}' = \{f' : f \in \mathcal{F}\}$ is uniformly bounded on compact subsets of Ω. For a closed disk $K \subseteq \Omega$ we have $|f'(z)| \leq M$ for all $z \in K$, $f' \in \mathcal{F}'$, and some constant M. Then for any two points z, $z' \in K$, integrating over a straight-line path from z to z' gives

$$|f(z) - f(z')| \le \int_{z'}^{z} |f'(\zeta)||d\zeta| \le M|z - z'|.$$

Hence, given $\varepsilon > 0$ and choosing $0 < \delta = \delta(\epsilon, K) < \frac{\epsilon}{M}$,

$$|f(z) - f(z')| < \varepsilon$$

whenever z, $z' \in K$, $|z - z'| < \delta$. Therefore $\mathcal{F}$ is equicontinuous on K, and a compactness argument completes the proof.

Note that the preceding proof depended on the functions of the family being analytic, as does another direct proof of Theorem 1.6.3 using the Cauchy Integral Formula.

The converse of the preceding theorem is false since the family

$$\mathcal{F} = \{f_n(z) = z + n : n = 1, 2, 3, \ldots\}$$

is equicontinuous in U, but not locally bounded.

Another important consequence of local boundedness is demonstrated in the following weak variant of the Vitali-Porter Theorem; the statement and proof of the latter are given in Chapter 2. The proof given here is completely elementary.

Theorem 1.6.4 *If $\{f_n\}$ is a locally bounded sequence of analytic functions on a domain Ω such that $\lim_{n \to \infty} f_n(z) = f(z) \in \mathbb{C}$, for each $z \in \Omega$, then $f_n \to f$ uniformly on compact subsets of Ω and f is analytic in Ω.*

Proof. We again conclude that $\mathcal{F}' = \{f_n'\}$ is uniformly bounded on compact subsets of Ω, so that given K a closed disk in Ω,

$$|f_n'(z)| \le M < \infty,$$

for all $z \in K$, $n = 1, 2, 3, \ldots$. Let $\varepsilon > 0$ and for each $z \in K$ define

$$U_z = \left\{\zeta \in \Omega : |z - \zeta| < \frac{\varepsilon}{3M}\right\}.$$

Then $\cup_{z \in K} U_z$ forms an open cover of K, so there are points z_1, z_2, ..., $z_k \in K$ such that $U_{z_1} \cup U_{z_2} \cup \ldots \cup U_{z_k} \supseteq K$. As each $z \in K$ belongs to some U_{z_i}, $|z - z_i| < \frac{\varepsilon}{3M}$ for all such z. Since $\{f_n\}$ converges at each $z \in \Omega$, there exists $n_0 \in \mathbb{N}$ such that $n, m \ge n_0$ implies

$$|f_n(z_i) - f_m(z_i)| < \frac{\varepsilon}{3}, \qquad 1 \le i \le k.$$

Then, for $n, m \ge n_0$, $z \in K$,

$$|f_n(z) - f_m(z)| \le |f_n(z) - f_n(z_i)| + |f_n(z_i) - f_m(z_i)| + |f_m(z_i) - f_m(z)|,$$

$1 \leq i \leq k$, and the middle term is less than $\frac{\varepsilon}{3}$ by the preceding remarks. To treat the first and third terms on the right, note that for $z \in U_{z_i}$

$$|f_n(z) - f_n(z_i)| \leq M|z - z_i| < \frac{\varepsilon}{3},$$

for $n = 1, 2, 3, \ldots$. We conclude that $|f_n(z) - f_m(z)| < \varepsilon$ for all $n, m \geq n_0$, $z \in K$, and so $f_n \to f$ uniformly on K, giving the analyticity of f. Again, by a compactness argument, the result follows.

1.7   Elliptic Functions

Elliptic integrals have a long and distinguished history dating back to the middle of the 18th century, but the investigation of the associated elliptic functions came more than 100 years later, initially in the works of Gauss, Abel, and Jacobi. There are two elliptic functions which play an important role in our discussion of normal families, namely the Weierstrass $\wp$-function and the elliptic modular function. Both of these functions have had a significant impact on the development of classical function theory, and the reader is referred to Segal [1981], Chapter 8, or Hille [1962] Vol. II, Chapter 13, for further details. We begin with a resumé of some basic features of elliptic functions.

Definition 1.7.1 *Let $f(z)$ be meromorphic in $\mathbb{C}$. We say that $f(z)$ is* **doubly periodic,** *or* **elliptic,** *if there are complex numbers ω, ω' such that*

$$\mathcal{I}m\left(\frac{\omega'}{\omega}\right) > 0,$$

and $f(z + \omega) = f(z)$, $f(z + \omega') = f(z)$, for $z \in \mathbb{C}$. We call ω, ω' **periods** *of f.*

If $f(z)$ has periods ω, ω', then clearly any linear combination $n\omega + m\omega'$, with n, m arbitrary integers, are also periods of $f(z)$. Hence we say that ω, ω' are *fundamental* or *primitive periods* if all others are integral multiples of ω, ω'. However, the pair of fundamental periods ω, ω' is not unique since it may be verified that

$$\omega_1 = a\omega + b\omega', \quad \omega_2 = c\omega + d\omega',$$

where $ad - bc = \pm 1$, is also a pair of fundamental periods.

Given fundamental periods ω, ω' of $f(z)$, then the points 0, ω, ω', $\omega + \omega'$ determine a parallelogram. This region, including that part of the boundary from 0 to ω and from 0 to ω'; and including 0 but not including ω or ω', is called a *fundamental parallelogram*. All the other congruent parallelograms with vertices determined by the points $n\omega + m\omega'$ are called *period parallelograms* (cf. Figure 1.3). Note that the complex plane $\mathbb{C}$ can be covered by a tesselation of period parallelograms.

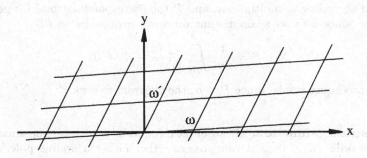

Figure 1.3

Thus, the range of values of an elliptic function is completely determined by its values in a period parallelogram. From this we conclude that an elliptic function which is analytic in a period parallelogram (that is, without poles) is constant by Liouville's theorem.

Definition 1.7.2 *The sum of the orders of the poles of an elliptic function within a period parallelogram is called the* **order** *of the function.*

The order must be positive if $f(z)$ is nonconstant, and indeed

Theorem 1.7.3 *The order of a nonconstant elliptic function f is at least 2.*

Proof. We first show that the sum of the residues of the poles of f in a period parallelogram is zero. To this end, let Π be a period parallelogram; it may be assumed that there are no poles of f on any of the sides of Π, for otherwise a suitable translation of Π may be considered. Then integrating over the sides of Π in the positive direction gives

$$\frac{1}{2\pi i} \int_{\partial \Pi} f(z)\, dz = 0$$

since the integrals over opposite sides of Π cancel each other by periodicity. As the integral represents the sum of the residues, the order of f cannot be equal to 1 and must be at least 2, as desired.

The order of an elliptic function determines how many times its values in $\mathbb{C}$ are taken. In fact,

Theorem 1.7.4 *A nonconstant function of order n takes every value in $\mathbb{C}$ exactly n times in a period parallelogram Π.*

Proof. Given $f(z)$ elliptic, nonconstant, note that the function

$$\frac{f'(z)}{f(z) - a}, \qquad a \in \mathbb{C},$$

is also elliptic. Letting N denote the number of zeros of $f(z) - a$ in Π counted according to multiplicity, and P the corresponding total for poles, we have (since we may again assume no zeros or poles lie on $\partial\Pi$)

$$N - P = \frac{1}{2\pi i} \int_{\partial\Pi} \frac{f'(z)}{f(z) - a} \, dz = 0$$

by the preceding result. Since $P = n$, the theorem follows.

Weierstrass $\wp$-function. If an elliptic function is to have the smallest possible order of 2, then it can possess either either a double pole with residue 0, or two simple poles having oppositely signed residues. The first case leads to the classical elliptic function of Weierstrass, and the second case to that of Jacobi. We pursue only the former.

Choose $\omega, \omega' \in \mathbb{C} - \{0\}$ with $\boldsymbol{Im}\left(\frac{\omega'}{\omega}\right) > 0$. Then the set of points $n\omega + m\omega'$, n, $m = 0, \pm 1, \pm 2, \ldots$, can be enumerated as the sequence $0 = \omega_0, \omega_1, \omega_2, \ldots, \omega_k, \ldots$. We define the *Weierstrass $\wp$-function* as

$$\wp(z) = \wp(z, \omega, \omega') = \frac{1}{z^2} + \sum_{k=1}^{\infty} \left[\frac{1}{(z - \omega_k)^2} - \frac{1}{\omega_k^2} \right]. \qquad (1.3)$$

It is not difficult to show (cf. Ahlfors [1979], pp. 272–273) that $\wp(z)$ is meromorphic in $\mathbb{C}$, having only double poles at the points ω_k, $k = 0, 1, 2, \ldots$

The series (1.3) at point $z \neq \omega_k$ converges absolutely so that the sum is independent of the order of its terms, and converges uniformly on any bounded domain not containing ω_k. Observe that $\wp(z)$ is an even function for

$$
\begin{aligned}
\wp(-z) &= \frac{1}{z^2} + \sum_{k=1}^{\infty} \left[\frac{1}{(z + \omega_k)^2} - \frac{1}{\omega_k^2} \right] \\
&= \frac{1}{z^2} + \sum_{k=1}^{\infty} \left[\frac{1}{\left(z - (-\omega_k)\right)^2} - \frac{1}{(-\omega_k)^2} \right] \\
&= \wp(z)
\end{aligned}
$$

since $\{-\omega_k\}$ is identical to the sequence $\{\omega_k\}$.

Furthermore, differentiating (1.3) term-by-term yields

$$\wp'(z) = \frac{-2}{z^3} - \sum_{k=1}^{\infty} \frac{2}{(z - \omega_k)^3} = -2 \sum_{k=0}^{\infty} \frac{1}{(z - \omega_k)^3}. \qquad (1.4)$$

Let us show that $\wp$ is indeed periodic. Reasoning as above, it follows that

$$\wp'(z+\omega) = \wp'(z) \quad \text{and} \quad \wp'(z+\omega') = \wp'(z).$$

Thus, if we integrate the equation $\wp'(z+\omega) - \wp'(z) = 0$, we obtain

$$\wp(z+\omega) - \wp(z) = c \quad \text{(constant)}.$$

Setting $z = -\frac{\omega}{2}$ in this equation gives

$$c = \wp\left(\frac{\omega}{2}\right) - \wp\left(-\frac{\omega}{2}\right) = 0$$

since $\wp(z)$ is even. We conclude that

$$\wp(z+\omega) = \wp(z),$$

and likewise, $\wp(z+\omega') = \wp(z)$, that is, $\wp(z)$ is doubly periodic and meromorphic, hence elliptic.

Thus, the numbers $\omega_k = n\omega + m\omega'$ are periods of $\wp(z)$, and in fact there are no others. For, if there were any others, then as $\wp(z)$ has a pole at $z = 0$, it would have to have poles at points different from the points ω_k, a contradiction. Hence, ω, ω' are fundamental periods for $\wp(z, \omega, \omega')$, and the Weierstrass $\wp$-function is elliptic of order 2.

From (1.4) we see that $\wp'(z)$ is an *odd* function, with ω, ω' fundamental periods. In view of the equations

$$\wp'\left(-\frac{\omega}{2}\right) = -\wp'\left(\frac{\omega}{2}\right), \quad \wp'\left(-\frac{\omega}{2}\right) = \wp'\left(\frac{\omega}{2}\right),$$

due to oddness and periodicity, respectively, we find

$$\wp'\left(\frac{\omega}{2}\right) = 0.$$

Similarly

$$\wp'\left(\frac{\omega'}{2}\right) = \wp'\left(\frac{\omega+\omega'}{2}\right) = 0.$$

Since $\wp'(z)$ has order 3, the points $\frac{\omega}{2}$, $\frac{\omega'}{2}$, $\frac{\omega+\omega'}{2}$ must be simple zeros of $\wp'(z)$ by Theorem 1.7.4, and the only zeros in the fundamental parallelogram. We infer that only at these three points are the roots of $\wp(z) = c$, $c \in \mathbb{C}$, of multiplicity 2. Write

$$\wp\left(\frac{\omega}{2}\right) = e_1, \quad \wp\left(\frac{\omega'}{2}\right) = e_2, \quad \wp\left(\frac{\omega+\omega'}{2}\right) = e_3.$$

This leads to an important feature of $\wp(z)$.

Theorem 1.7.5 *The equation $\wp(z) = c$, $c \in \widehat{\mathbb{C}}$, has exactly one double root, for $c \in \{e_1, e_2, e_3, \infty\}$, and no multiple roots for any other value of c.*

That the numbers e_1, e_2, e_3 are distinct is clear from the fact that if any two were the same, that value would be assumed four times, contradicting (in view of Theorem 1.7.4) the fact that the order of $\wp(z)$ is 2. If c is a value distinct from e_1, e_2, e_3, ∞, then $\wp(z) = c$ has two different simple roots in the fundamental parallelogram.

Elliptic Modular Function. While this function can be constructed directly from the quantities e_1, e_2, e_3 of the Weierstrass $\wp$-function (cf., e.g., Ahlfors [1979], p. 278), we pursue an alternative, but equivalent, approach.

In the open unit disk U, consider a domain Ω_0 which is the interior of a regular hyperbolic triangle whose sides are arcs of circles which meet $|z| = 1$ orthogonally at the vertices, each of which has zero angle (cf. Figure 1.4). There is a univalent analytic mapping, due to Schwarz, of the upper half-plane onto Ω_0, and the inverse of this mapping we denote by $\mu(z)$. The three boundary arcs of Ω_0 are mapped by μ one-to-one continuously onto the extended real line, and it can be arranged so that the three vertices are mapped to 0, 1, ∞, respectively. We proceed to extend μ to all of U.

First, reflect Ω_0 in U about each of its sides to obtain the domains Ω_1, Ω_2, Ω_3. Since $\mu(z)$ assumes real values on the three boundary arcs of Ω_0, the Schwarz Reflection Principle permits $\mu(z)$ to be extended analytically to Ω_1, Ω_2, Ω_3, mapping each of these regions univalently and analytically onto $\mathcal{I}m(\zeta) < 0$. Then $\mu(z)$ is analytic in $\Omega' = \Omega_0 \cup \Omega_1 \cup \Omega_2 \cup \Omega_3$; however, it is no longer univalent there. The six boundary arcs of Ω' are orthogonal to $|z| = 1$, forming vertices of zero angle (since reflection is a conformal mapping) and μ is again real on these arcs. Reflecting Ω_i, $i = 1, 2, 3$, across each of their corresponding sides allows one to extend $\mu(z)$ into the six new (shaded) regions (cf. Figure 1.4), each of which maps analytically and univalently onto the upper half-plane.

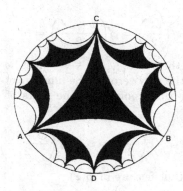

Figure 1.4

Continuing in this fashion, a *triangulation* of the entire disk U ensues. Furthermore, $\mu(z)$ is analytic in U and assumes every value except 0, 1, ∞. For any $\zeta = \mu(z)$ in the upper half-plane, z will be in one of the shaded triangular regions. The action of two consecutive reflections carries $z \rightarrow z'$, with $\zeta = \mu(z) = \mu(z')$. In other words, $\mu(z)$ is invariant under an even number of reflections of the argument. The group consisting of products of an even number of reflections, as above, is an example of a *triangle group*, specifically, the *congruence subgroup* $\Gamma(2)$, and $\mu(z)$ is an *automorphic function* with respect to this subgroup of the *modular group* (cf. §5.5 for further details). A fundamental region consists of the union of a shaded and unshaded triangle. The function $\mu(z)$ is known as the *elliptic modular function* although, in general, any automorphism with respect to a subgroup of the modular group is called a *modular function*.

The inverse function

$$\nu(w) = \mu^{-1}(w)$$

is multiple-valued, yet each function element can be analytically continued along any curve not passing through 0, 1, and ∞. We distinguish the *principal value* of $\nu(w)$ as the one taken in the quadrilateral $ADBC$ not including the sides AD and DB.

1.8   Nevanlinna Theory

The theory of the distribution of values of a meromorphic function, developed by R. Nevanlinna and L. Ahlfors was one of the most outstanding achievements in function theory this century. Since its development in the 1920s (cf. Nevanlinna [1925, 1970]) the former especially has become an indispensable tool in the study of normal families. A brief account is presented here; for a more comprehensive treatment, the reader is referred to the excellent monograph of Hayman [1964].

The starting point of the theory is the Poisson-Jensen formula which can be deduced from the Poisson formula given in §5.4.

Theorem 1.8.1 *Let $f(z)$ be meromorphic in $|z| \leq r < \infty$ and have zeros $a_1, a_2, \ldots, a_N$, and poles $b_1, b_2, \ldots, b_M$ in $|z| < r$, repeated according to multiplicity. If $f(z) \neq 0$, ∞ and $z = \rho e^{i\theta}$, $0 \leq \rho < r$, then*

$$
\begin{aligned}
\log|f(z)| = {} & \frac{1}{2\pi} \int_0^{2\pi} \log|f(re^{i\phi})| \frac{r^2 - \rho^2}{r^2 - 2r\rho\cos(\theta - \phi) + \rho^2} \, d\phi \\
& + \sum_{i=1}^{N} \log\left|\frac{r(z - a_i)}{r^2 - \overline{a}_i z}\right| - \sum_{j=1}^{M} \log\left|\frac{r(z - b_j)}{r^2 - \overline{b}_j z}\right|.
\end{aligned}
$$

When $z = 0$, this yields *Jensen's formula*

$$\log|f(0)| = \frac{1}{2\pi}\int_0^{2\pi}\log|f(re^{i\phi})|\,d\phi - \sum_{i=1}^{N}\log\frac{r}{|a_i|} + \sum_{j=1}^{M}\log\frac{r}{|b_j|}. \quad (1.5)$$

If $f(z)$ has a zero or a pole at $z = 0$, then both sides of Jensen's formula are infinite. However, with a suitable modification of $f(z)$ these two cases lead to a similar but more cumbersome version of Jensen's formula. Bearing this is mind, our analysis in the sequel will be carried out for (1.5) as given and the terms finite.

Define $n(t, 0)$ and $n(t, \infty)$ as the number of zeros and poles in $|z| \leq t$, respectively, counted according to multiplicity. If $f(z)$ is as in Theorem 1.8.1 and $t_1, t_2, \ldots, t_N$ are the moduli of the zeros $a_1, a_2, \ldots, a_N$, then

$$\sum_{i=1}^{N}\log\frac{r}{|a_i|} = \sum_{i=1}^{N}\log\frac{r}{t_i} = \int_0^{2\pi}\log\frac{r}{t}\,dn(t, 0)$$

$$= \left[\log\frac{r}{t}n(t, 0)\right]_0^r - \int_0^{2\pi}n(t, 0)\,d\log\frac{r}{t} = \int_0^r\frac{n(t, 0)}{t}\,dt,$$

where we have applied integration by parts to the Stieltjes integral. Similarly,

$$\sum_{j=1}^{M}\log\frac{r}{|b_j|} = \int_0^r\frac{n(t, \infty)}{t}\,dt.$$

Denote

$$N(r, f) = N(r, \infty) = \int_0^r\frac{n(t, \infty)}{t}\,dt = \sum_{j=1}^{M}\log\frac{r}{|b_j|},$$

$$N\left(r, \frac{1}{f}\right) = \int_0^r\frac{n(t, 0)}{t}\,dt = \sum_{i=1}^{N}\log\frac{r}{|a_i|}.$$

Then (1.5) becomes

$$\frac{1}{2\pi}\int_0^{2\pi}\log|f(re^{i\phi})|\,d\phi = \log|f(0)| + N\left(r, \frac{1}{f}\right) - N(r, f).$$

Setting

$$\log^+ x = \max(\log x, 0) \qquad (\log^+ 0 = 0),$$

so that $\log x = \log^+ x - \log^+ \frac{1}{x}$ for $x > 0$, we then obtain

$$\frac{1}{2\pi}\int_0^{2\pi}\log|f(re^{i\phi})|\,d\phi = \frac{1}{2\pi}\int_0^{2\pi}\log^+|f(re^{i\phi})|\,d\phi$$

$$- \frac{1}{2\pi}\int_0^{2\pi}\log^+\frac{1}{|f(re^{i\phi})|}\,d\phi.$$

If we denote the *proximity function* (*Schmiegungsfunktion*) by

$$m(r, f) = m(r, \infty) = \frac{1}{2\pi} \int_0^{2\pi} \log^+ |f(re^{i\phi})|\, d\phi,$$

then (1.5) now reads

$$m(r, f) - m\left(r, \frac{1}{f}\right) = \log |f(0)| + N\left(r, \frac{1}{f}\right) - N(r, f),$$

i.e.,

$$m(r, f) + N(r, f) = m\left(r, \frac{1}{f}\right) + N\left(r, \frac{1}{f}\right) + \log |f(0)|.$$

The *characteristic function* of $f(z)$ is then given by the quantity

$$T(r, f) = m(r, f) + N(r, f).$$

As a consequence, Jensen's formula can be written as

$$T(r, f) = T\left(r, \frac{1}{f}\right) + \log |f(0)|. \tag{1.6}$$

At this juncture we require some elementary inequalities:

$$\log^+\left(\sum_{k=1}^p x_k\right) \leq \sum_{k=1}^p \log^+ x_k + \log p$$

and

$$\log^+(x_1 \cdot x_2 \cdots x_p) \leq \sum_{k=1}^p \log^+ x_k.$$

These yield for meromorphic functions $f_1, f_2, \ldots, f_p$,

$$m\left(r, \sum_{k=1}^p f_k\right) \leq \sum_{k=1}^p m(r, f_k) + \log p$$

and

$$m(r, f_1 \cdot f_2 \cdots f_p) \leq \sum_{k=1}^p m(r, f_k).$$

Furthermore, the order of a pole at a point z_0 of the sum or product of the functions $f_k(z)$ is certainly less than or equal to the sum of the orders of the poles of the $f_k(z)$ at z_0. This leads to

$$N\left(r, \sum_{k=1}^p f_k\right) \leq \sum_{k=1}^p N(r, f_k)$$

and

$$N(r, f_1 \cdot f_2 \cdots f_p) \le \sum_{k=1}^{p} N(r, f_k).$$

Thus, for the proximity function,

$$T\left(r, \sum_{k=1}^{p} f_k\right) \le \sum_{k=1}^{p} T(r, f_k) + \log p$$

and

$$T(r, f_1 \cdot f_2 \cdots f_p) \le \sum_{k=1}^{p} T(r, f_k).$$

Whence for $a \in \mathbb{C}$,

$$T(r, f - a) \le T(r, f) + \log^+ |a| + \log 2,$$
$$T(r, f) \le T(r, f - a) + \log^+ |a| + \log 2,$$

implying

$$|T(r, f) - T(r, f - a)| \le \log^+ |a| + \log 2. \qquad (1.7)$$

Introducing the notation

$$m(r, a) = m\left(r, \frac{1}{f - a}\right), \quad N(r, a) = N\left(r, \frac{1}{f - a}\right),$$

we can establish the

First Fundamental Theorem *If $f(z)$ is meromorphic in $|z| < R \le \infty$, then for any $a \in \mathbb{C}$ and $0 < r < R$,*

$$m(r, a) + N(r, a) = T\left(r, \frac{1}{f - a}\right) = T(r, f) - \log |f(0) - a| + \varepsilon(r, a),$$

where $\varepsilon(r, a) \le \log^+ |a| + \log 2$.

In fact, Jensen's formula (1.5) in terms of $f - a$ is

$$T(r, f - a) = T\left(r, \frac{1}{f - a}\right) + \log |f(0) - a|.$$

The theorem then follows from an application of (1.7).

Observe that the term $N(r, a)$ is a weighted measure of the density of the a-points of $f(z)$ in $|z| < r$, whereas $m(r, a)$ is in some sense an inverse measure of the mean deviation of $f(z)$ from a on $|z| = r$. The theorem states that the sum of these two quantities remains the same up to a bounded term, specifically

$$m(r, a) + N(r, a) = T(r, f) + O(1) \qquad \text{as} \qquad r \to R,$$

where a can be finite or infinite.

We shall require a growth estimate of the characteristic function $T(r, f)$. When $f(z)$ is analytic, there is the following inequality.

Theorem 1.8.2 *Let $f(z)$ be analytic in $|z| \leq R$, and define*

$$M(r, f) = \max_{|z|=r} |f(z)|.$$

Then

$$T(r, f) \leq \log^+ M(r, f) \leq \frac{R+r}{R-r} T(R, f), \qquad 0 \leq r < R.$$

Proof. Since $N(r, f) = 0$,

$$T(r, f) = m(r, f) = \frac{1}{2\pi} \int_0^{2\pi} \log^+ |f(re^{i\phi})|\, d\phi \leq \log^+ M(r, f).$$

Furthermore, if $z = re^{i\theta}$, $0 \leq r < R$, and if $a_1, a_2, \ldots, a_N$ denote the zeros of $f(z)$ in $|z| < R$, then by Theorem 1.8.1,

$$
\begin{aligned}
\log |f(z)| &= \frac{1}{2\pi} \int_0^{2\pi} \log |f(Re^{i\phi})| \frac{R^2 - r^2}{R^2 - 2Rr\cos(\theta - \phi) + r^2}\, d\phi \\
&\quad + \sum_{i=1}^N \log \left| \frac{R(z - a_i)}{R^2 - \bar{a}_i z} \right| \\
&\leq \frac{1}{2\pi} \int_0^{2\pi} \log^+ |f(Re^{i\phi})| \frac{R^2 - r^2}{R^2 - 2Rr\cos(\theta - \phi) + r^2}\, d\phi \\
&\leq \frac{R+r}{R-r} m(R, f) \qquad (f(z) \neq 0).
\end{aligned}
$$

Consequently,

$$\log^+ |f(z)| \leq \frac{R+r}{R-r} T(R, f),$$

and the result follows.

Moreover, Cartan [1929] gave the following useful identity for the characteristic function.

Theorem 1.8.3 *If $f(z)$ is meromorphic in $|z| < R$, then*

$$T(r, f) = \frac{1}{2\pi} \int_0^{2\pi} N(r, e^{i\theta})\, d\theta + \log^+ |f(0)|, \qquad 0 < r < R.$$

From this we conclude (cf. also Nevanlinna [1970], p. 175)

Corollary 1.8.4 $T(r, f)$ *is an increasing convex function of* $\log r$, $0 < r < R$.

This follows from the fact that $N(r, e^{i\theta})$ has this growth property.

Ahlfors-Shimizu Characteristic. Another approach to the theory of meromorphic functions was taken independently by Ahlfors [1929] and Shimizu [1929]. For a meromorphic function $f(z)$ let $S(r)$ be given by (1.2). Define the *Ahlfors-Shimizu characteristic* $T_0(r, f)$ by

$$T_0(r, f) = \int_0^r \frac{S(t)}{t} \, dt.$$

The relation with the Nevanlinna characteristic $T(r, f)$ is given by

$$|T(r, f) - T_0(r, f) - \log^+ |f(0)|| \le \frac{1}{2} \log 2,$$

that is, the two characteristic functions differ by a bounded term, and for this reason are used more or less interchangeably.

Suppose that $f(z)$ is nonconstant meromorphic in $\mathbb{C}$, so that $S(r) > 0$ for all $r > 0$. For an arbitrary $r_0 > 0$, and $r > r_0$, $S(r) > S(r_0)$, implying

$$T_0(r, f) = \int_0^r \frac{S(t)}{t} \, dt > \int_{r_0}^r \frac{S(t)}{t} \, dt > S(r_0) \log \frac{r}{r_0}, \qquad (1.8)$$

a fortiori

$$\lim_{r \to \infty} \frac{T_0(r, f)}{\log r} > 0.$$

We have shown

Theorem 1.8.5 *If the characteristic* $T_0(r, f)$ *of a function* $f(z)$ *that is meromorphic in* $\mathbb{C}$ *satisfies*

$$\lim_{r \to \infty} \frac{T_0(r, f)}{\log r} = 0,$$

then $f(z) \equiv$ *constant, in particular, if* $T_0(r, f)$ *is bounded for all* $r > 0$, *then we conclude* $f(z) \equiv$ *constant.*

On the other hand, for a meromorphic function in $|z| < R < \infty$, the characteristic may be bounded. These form an important class known as functions of *bounded type* (cf. Nevanlinna [1970], Chapter 7; also Theorem 3.6.8, and §2.3 Example 7, for the analytic counterpart).

We next give a basic inequality whose proof can be found in Hayman [1964], pp. 31–34. It is one formulation of the Second Fundamental Theorem and plays an important role in Chapter 4.

Fundamental Inequality *Let* $f(z)$ *be a nonconstant meromorphic function in* $|z| \le r$. *Suppose that* $a_1, a_2, \ldots a_q$, $q \ge 2$, *are distinct complex* *numbers* ($\neq \infty$) *satisfying* $|a_i - a_j| \geq \delta > 0$ *for* $1 \leq i < j \leq q$. *If* $f(0) \neq 0$, ∞, *and* $f'(0) \neq 0$, *then*

$$m(r, f) + \sum_{i=1}^{q} m(r, a_i) \leq 2T(r, f) - N_1(r) + S(r),$$

where

$$0 < N_1(r) = N\left(r, \frac{1}{f'}\right) + 2N(r, f) - N(r, f')$$

and

$$S(r) = m\left(r, \frac{f'}{f}\right) + m\left(r, \sum_{i=1}^{q} \frac{f'}{f - a_i}\right) - q\log^+ \frac{3q}{\delta} + \log 2 + \log \frac{1}{|f'(0)|}.$$

The term N_1 measures the number of multiple points of $f(z)$, whereas $S(r)$ plays the role of a negligible error term.

We now wish to make some previous definitions more specific. Let $f(z)$ be meromorphic in $|z| < R$, denote by $n(t, a)$ the number of roots of $f(z) - a = 0$ in $|z| \leq t$, counted according to multiplicity, and let $\bar{n}(t, a)$ be the number of distinct roots of $f(z) - a = 0$ in $|z| \leq t$. Define

$$N(r, a) = N(r, a, f) = \int_0^r \frac{n(t, a) - n(0, a)}{t} \, dt + n(0, a) \log r,$$

$$\bar{N}(r, a) = \bar{N}(r, a, f) = \int_0^r \frac{\bar{n}(t, a) - \bar{n}(0, a)}{t} \, dt + \bar{n}(0, a) \log r.$$

Note that $N(r, a)$ is the same as before if $f(0) \neq a$, whereas if $n(0, a) \neq 0$, then for some sufficiently small $r_1 > 0$, $n(t, a) - n(0, a) = 0$ for all $t \in [0, r_1]$. Then

$$N(r, a) = \int_{r_1}^r \frac{n(t, a)}{t} \, dt + n(0, a) \log r_1.$$

We write $m(r, f) = m(r, \infty)$, $N(r, f) = N(r, \infty)$, and $T(r, f) = m(r, f) + N(r, f)$ as previously. Then by the First Fundamental Theorem

$$m(r, a) + N(r, a) = T(r, f) + O(1), \qquad r \to R, \ a \in \widehat{\mathbb{C}}.$$

Set $n(t, f) = n(t, \infty)$, and note that if z_0 is a pole of order p of $f(z)$, then z_0 is a pole of order $p + 1$ of $f'(z)$. Hence

$$n(t, f') = n(t, f) + \bar{n}(t, f)$$

and

$$N(r, f') = N(r, f) + \bar{N}(r, f).$$

Likewise

$$N\left(r, \frac{1}{f'}\right) = N\left(r, \frac{1}{f}\right) - \bar{N}\left(r, \frac{1}{f}\right).$$

Let $f(z)$ be meromorphic in $|z| < R \leq \infty$, and assume $T(r, f) \to \infty$ as $r \to R$. Define

$$
\begin{aligned}
\delta(a) &= \delta(a, f) = \varliminf_{r \to R} \frac{m(r, a)}{T(r, f)} = 1 - \varlimsup_{r \to R} \frac{N(r, a)}{T(r, f)}, \\
\Theta(a) &= \Theta(a, f) = 1 - \varlimsup_{r \to R} \frac{\bar{N}(r, a)}{T(r, f)}, \\
\theta(a) &= \theta(a, f) = \varliminf_{r \to R} \frac{N(r, a) - \bar{N}(r, a)}{T(r, f)}.
\end{aligned}
$$

These quantities satisfy $0 \leq \psi(a) \leq 1$, for $\psi = \delta, \Theta, \theta$. The term $\delta(a)$ is the *deficiency* of the value a, and $\theta(a)$ is the *index of multiplicity* or *ramification index*.

Then we have the celebrated

Second Fundamental Theorem *The set of values $a \in \widehat{\mathbb{C}}$ for which $\Theta(a) > 0$ is countable, and summing over these values*

$$
\sum_a \big(\delta(a) + \theta(a)\big) \leq \sum_a \Theta(a) \leq 2, \tag{1.9}
$$

provided that either $R = \infty$ and $f(z) \not\equiv$ constant, or $R < \infty$ and

$$
\varlimsup_{r \to R} \frac{T(r, f)}{\log \frac{1}{R-r}} = \infty.
$$

Example 1.8.6 (Picard's First Theorem) Suppose $f(z)$ is nonconstant meromorphic in $\mathbb{C}$ and $f(z) \neq a \in \widehat{\mathbb{C}}$. Then $N(r, a) = 0$ and $\delta(a) = 1$. By (1.9), since $\theta(a) \geq 0$ there can be at most two omitted values (cf. also §2.8).

Example 1.8.7 Let $f(z)$ be nonconstant meromorphic in $\mathbb{C}$ such that all the roots of $f(z) = a_\nu$ have multiplicity at least $m_\nu \geq 2$. Then

$$
\bar{N}(r, a_\nu) \leq \frac{1}{m_\nu} N(r, a_\nu) \leq \frac{1}{m_\nu} T(r, f) + O(1),
$$

which implies $\Theta(a_\nu) \geq 1 - \frac{1}{m_\nu}$. By (1.9),

$$
\sum_\nu \left(1 - \frac{1}{m_\nu}\right) \leq 2, \tag{1.10}
$$

and since $\Theta(a_\nu) \geq \frac{1}{2}$, at most four values a_ν can exist. The Weierstrass $\wp$-function (§1.7) has exactly four values each of multiplicity two.

Corollary 1.8.8 *Let $f(z)$ be meromorphic in $\mathbb{C}$ with the following properties: (i) all the zeros of $f(z)$ have multiplicity $\geq h$, (ii) all the poles have multiplicity $\geq k$, (iii) all the zeros of $f(z) - 1$ have multiplicity $\geq \ell$. If*

$$\frac{1}{h} + \frac{1}{k} + \frac{1}{\ell} < 1,$$

then $f(z) \equiv$ constant.

For, if $f(z)$ were nonconstant, then

$$\left(1 - \frac{1}{h}\right) + \left(1 - \frac{1}{k}\right) + \left(1 - \frac{1}{\ell}\right) > 2,$$

which would contradict (1.10).

1.9   Ahlfors Theory of Covering Surfaces

A more geometric approach to value distribution theory than the one expounded by Nevanlinna was introduced by Ahlfors [1935]. The Ahlfors theory, while having many parallel features in common with the Nevanlinna theory, has proved a powerful tool in its own right and has led to such new insights as the connection between Picard's theorem and the Euler characteristic of topology.

To begin with, suppose that $f(z)$ is meromorphic in $|z| \leq r$, let D be a domain on the Riemann sphere, and let $I(r, D)$ denote the area of the image $f(\{|z| < r\})$ which lies over D (with regard to multiplicity), with $I_0(D)$ denoting the area of D. In this section, all domains on the Riemann sphere are taken to be Jordan domains each of which is bounded by a *sectionally analytic (s.a.) Jordan curve*. An s.a. Jordan curve is a simple closed curve consisting of a finite union of regular analytic arcs $P_1P_2, P_2P_3, \ldots, P_nP_1$, such that at the points P_i $(i = 1, \ldots, n)$ where two arcs meet, they do not form a cusp (cf. Hayman [1964], p. 126). Setting $S(r, D) = \frac{I(r,D)}{I_0(D)}$, we state the

First Fundamental Theorem *There exists a constant $h_1 = h_1(D)$ such that*

$$|S(r) - S(r, D)| \leq h_1 L(r).$$

Here $L(r)$ is the length of $f(\{|z| = r\})$ on Σ and $S(r) = \frac{1}{\pi} \cdot$ area of $f(\{|z| < r\})$ on Σ, with due regard to multiplicity, as given by (1.1) and (1.2), respectively.

Furthermore, suppose Δ is a subdomain of $|z| < r$, with $\overline{\Delta} \cap \{|z| = r\} = \emptyset$, which is mapped by $f(z)$ in a p-to-one fashion onto D. Then Δ is an *island over D of multiplicity p*, and in this instance, such as island contributes the quantity p to $S(r, D)$. If $p = 1$, then D is a *simple island*.

Second Fundamental Theorem *Let $D_1, \ldots, D_q$, $q \geq 3$, be Jordan domains on the w-sphere having disjoint closures. Then there exists a constant h_2 depending only on the domains D_j such that*

$$\sum_{j=1}^{q} \left(S(r) - \bar{n}(r, D_j) \right) \leq 2S(r) + h_2 L(r), \tag{1.11}$$

where $\bar{n}(r, D)$ is the total number of distinct islands over D in $|z| < r$ without regard to multiplicity.

Suppose $w = f(z)$ is meromorphic and nonconstant in $|z| < R \leq \infty$. If there is an increasing sequence $r_n \to R$ such that

$$\lim_{n \to \infty} \frac{L(r_n)}{S(r_n)} = 0, \tag{1.12}$$

then we say that the Riemann surface of $f(z)$ is *regularly exhaustible*. This will be the case, for example, if $R = \infty$, or $\overline{\lim}_{r \to R} S(r)(R - r) = \infty$ when $R < \infty$. The assumption of (1.12) gives many of the striking results of the Ahlfors theory, and in this case, $L(r)$ may be viewed as a negligible error term in both fundamental theorems.

As a consequence of the Second Fundamental Theorem we have a remarkable generalization of Bloch's theorem (cf. §4.3).

Ahlfors Five Islands Theorem *Let $w = f(z)$ be nonconstant meromorphic in $|z| < R \leq \infty$ and have a regularly exhaustible Riemann surface. If $D_1, \ldots, D_5$ are Jordan domains on the w-sphere having disjoint closures, then $f(z)$ has a simple island Δ over at least one of the domains D_j, that is to say, $f|\Delta$ is a one-to-one mapping of Δ onto D_j.*

Proof. First suppose that each D_j has an island, or islands, over it and each such island has multiplicity ≥ 2. Then from the First Fundmental Theorem

$$\bar{n}(r, D_j) \leq \frac{1}{2} S(r, D_j) \leq \frac{1}{2} S(r) + h_j L(r), \qquad 0 < r < R, j = 1, \ldots, 5.$$

When this is applied to (1.11), the inequality

$$S(r) < hL(r) \tag{1.13}$$

obtains, and this violates condition (1.12). If there is no island over some D_j, we similarly arrive at (1.13), proving the theorem.

Consider the Weierstrass $\wp$-function, which has no simple roots to the equations $\wp(z) = e_1, e_2, e_3, \infty$. This results in there being four disjoint domains on Σ over which there are no simple islands. Thus, the number 5 in the preceding is best possible. For a precursor to the Five Islands Theorem, see Bloch [1926b].

Finally, suppose that $w = f(z)$ is meromorphic in $|z| < R \leq \infty$ with a regularly exhaustible Riemann surface and, as above, let $D_1, \ldots, D_q$, $q \geq 3$, be Jordan domains on the w-sphere with disjoint closures. If every island over D_ν has multiplicity $\geq \mu_\nu$, then

$$\sum_{\nu=1}^{q} \left(1 - \frac{1}{\mu_\nu} \right) \leq 2, \tag{1.14}$$

which can be readily deduced from the Second Fundamental Theorem (cf. Hayman [1964], p. 148). This inequality is the geometric equivalent of the one already encountered in the Nevanlinna theory (1.10). The Five Islands Theorem follows immediately by taking $q = 5$, $\mu_\nu = 2$. If $f(z)$ is nonconstant entire (or analytic in $|z| < R$), and D_1, D_2, D_3 are *bounded* Jordan domains on the w-sphere with disjoint closures, $\infty \notin D_j$, $j = 1, 2, 3$, append an additional such domain D_4 disjoint from the others and containing ∞. Since $f(z)$ can have no islands over D_4, we can choose $\mu_4 = \infty$ and try $\mu_1 = \mu_2 = \mu_3 = 2$. As this contradicts (1.14) we infer that $f(z)$ has a simple island over one of the three given domains. This is the *Three Islands Theorem*. It is evident that it may be derived directly from (1.11).

2

Analytic Functions

It is certainly possible to treat normal families of analytic functions under the heading of meromorphic functions by considering the convergence only in terms of the spherical metric. We have considered analytic and meromorphic functions separately, in part for the sake of simplicity, partly because this was the approach historically, and partly to illustrate some of the difference between the two classes of functions. Results on normal families of analytic functions pertaining to the so-called "Bloch Principle" have been reserved for Chapter 4.

Perhaps the most far-reaching result in the theory of normal families is the Fundamental Normality Test (FNT) (*Critère fondamental*) of Montel. From it flow effortlessly the classical theorems of Picard, Schottky, and Julia, as well as various covering theorems. In addition to presenting a variation of Montel's orginal proof of the FNT based on the elliptic modular function, four further proofs are given later in the text.

2.1 Normality

Montel's work on normal families was intitiated with his 1907 paper on sequences of functions, although the term *normal* did not appear until his 1911 *Comptes Rendus* announcement of his seminal work published a year later. Montel's final word on the subject was written in 1946.

Definition 2.1.1 *A family $\mathcal{F}$ of analytic functions on a domain $\Omega \subseteq \mathbb{C}$ is **normal** in Ω if every sequence of functions $\{f_n\} \subseteq \mathcal{F}$ contains either a subsequence which converges to a limit function $f \not\equiv \infty$ uniformly on each compact subset of Ω, or a subsequence which converges uniformly to ∞ on each compact subset.*

In the former case, the limit function is an analytic function by the Weierstrass Theorem (§1.4). In the latter, we mean that for each compact subset $K \subseteq \Omega$ and constant $M > 0$, $|f_n(z)| > M$ for all $z \in K$, provided n (which may depend on K and M) is sufficiently large. For a given sequence $\{f_n\}$, both alternatives may occur.

Convergence of a subsequence to ∞ is occasionally omitted by some authors from the definition of normality. However, the above definition permits a more comprehensive treatment of the subject and greater compatibility with the normality of families of meromorphic functions in Chapter 3.

The family $\mathcal{F}$ is said to be *normal at a point* $z_0 \in \Omega$ if it is normal in some neighbourhood to z_0.

Theorem 2.1.2 *A family of analytic functions* $\mathcal{F}$ *is normal in a domain* Ω *if and only if* $\mathcal{F}$ *is normal at each point of* Ω.

Proof. It is evident that if $\mathcal{F}$ is normal in Ω, it is normal at each of its points.

Conversely, suppose that $\mathcal{F}$ is normal at each $z \in \Omega$. Choose a countable dense subset $\{z_n\}$ of Ω, say $z_n = x_n + iy_n$, where x_n and y_n are rational. Denote by $D(z_n; r_n)$ the largest disk about z_n in which $\mathcal{F}$ is normal. Note that if $z_{n_k} \to \zeta \in \Omega$, then $\zeta \in D(z_{n_k}; r_{n_k})$ for some k sufficiently large since $r_{n_k} \to 0$ only if $\zeta \in \partial\Omega$. We conclude that $\cup_{n=1}^{\infty} D(z_n; r_n)$ covers Ω. For any sequence $\{f_n\} \subseteq \mathcal{F}$, one can extract a convergent subsequence $\{f_{n_k}^{(1)}\}$ which converges uniformly in $D(z_1; r_1/2)$ to an analytic function or to ∞. The sequence $\{f_{n_k}^{(1)}\}$ in turn has a subsequence $\{f_{n_k}^{(2)}\}$ which converges uniformly in $D(z_2; r_2/2)$ and $D(z_1; r_1/2)$. Continuing in this fashion, it follows that the diagonal sequence $\{f_{n_k}^{(k)}\}$ converges uniformly in $D(z_n; r_n/2)$, $n = 1, 2, 3, \ldots$, to either an analytic function or to ∞. This dichotomy divides the points $z \in \Omega$ into two classes, say Ω_0; Ω_∞, which are open sets satisfying $\Omega_0 \cap \Omega_\infty = \emptyset$, $\Omega_0 \cup \Omega_\infty = \Omega$. The connectedness of Ω means that either $\Omega_0 = \Omega$ or $\Omega_\infty = \Omega$. If K is a compact subset of Ω, the family of open disks $\{D(z_n; r_n/2)\}$ forms an open cover of K, and since K can be covered by a finite number of such disks, the result follows.

We now stipulate that a family of analytic functions $\mathcal{F}$ is *normal at* ∞ if the corresponding family $\mathcal{G} = \{g : g(z) = f(\frac{1}{z})\}$ is normal at $z = 0$. Therefore, $\mathcal{F}$ normal at ∞ means it is normal in some neighbourhood of ∞, $\Delta(\infty; R) = \{z : |z| > R\} \cup \{\infty\}$, $R > 0$. This leads us to define $\mathcal{F}$ to be *normal* in a domain Ω which lies in $\widehat{\mathbb{C}}$ and contains the point at infinity, if $\mathcal{F}$ is normal in $\Omega - \{\infty\}$ in the usual sense, as well as normal at ∞ in the above sense. This definition is equivalent to letting Ω belong to $\widehat{\mathbb{C}}$ in Definition 2.1.1.

In view of Theorem 2.1.2, if a family $\mathcal{F}$ is not normal in a domain Ω, there is at least one point in Ω at which $\mathcal{F}$ is not normal. Such a point is called an *irregular point* and is the point of departure for the study of *quasi-normal families* (see Appendix). Such points are also important in the study of complex dynamical systems (§5.2), where they form what are known as *Julia sets.*

Theorem 2.1.2 also permits us to establish the normality of a family $\mathcal{F}$ by demonstrating that $\mathcal{F}$ has this property locally. A simple, but important, example in the theory of normal (and quasi-normal) families is

Example 2.1.3 Let $\mathcal{F} = \{f_n(z) = nz : n = 1, 2, 3, \ldots\}$. Then $f_n(0) \to 0$, but $f_n(z) \to \infty$ as $n \to \infty$ for $z \neq 0$. Hence $\mathcal{F}$ cannot be normal in any domain containing the origin.

To those of us in New Zealand (abbreviated NZ) the preceding example could be construed as somewhat of an embarrassment. Fortunately, the family is at least quasi-normal (refer to the Appendix).

Observe that the normality of a family $\mathcal{F}$ of analytic functions on a domain Ω is *conformally invariant*. That is, if $\phi(\zeta) = z$ is a one-to-one conformal mapping of Ω' onto Ω, then the family $\mathcal{G} = \{g = f \circ \phi : f \in \mathcal{F}\}$ is normal in Ω' if and only if $\mathcal{F}$ is normal in Ω. This observation, coupled with Theorem 2.1.2, allows us to confine our considerations of normality to the unit disk U, when it is convenient to do so.

2.2 Montel's Theorem

One of the progenitors of Montel's theory of normal families was the following theorem of Ascoli [1883] and Arzelà [1889, 1895, 1899], in various formulations commonly known as the Arzelà-Ascoli Theorem. The roots of Montel's theory lie within this theorem.

Theorem 2.2.1 *If a sequence $\{f_n\}$ of continuous functions converges uniformly on a compact set K to a limit function $f (\not\equiv \infty)$, then $\{f_n\}$ is equicontinuous on K, and f is continuous. Conversely, if $\{f_n\}$ is equicontinuous and locally bounded on Ω, then a subsequence can be extracted from $\{f_n\}$ which converges locally uniformly in Ω to a (continuous) limit function f.*

The first implication of the theorem has already been encounted in Proposition 1.6.2. Regarding the second implication, Montel observed in his 1907 paper that for a family of *analytic* functions, local boundedness implies equicontinuity (Theorem 1.6.3). This leads us to

Montel's Theorem *If $\mathcal{F}$ is a locally bounded family of analytic functions on a domain Ω, then $\mathcal{F}$ is a normal family in Ω.*

Proof. As in Theorem 2.1.2, choose a countable dense subset $\{z_n\}$ in Ω. Take any sequence $\{f_n\} \subseteq \mathcal{F}$ and consider the sequence of complex numbers $\{f_n(z_1)\}$. By hypothesis, $|f_n(z_1)| < M$ for some constant M and $n = 1, 2, 3, \ldots$. This bounded sequence of complex numbers has a convergent subsequence by the Bolzano-Weierstrass property, say

$$f_{n_1}^{(1)}(z_1), f_{n_2}^{(1)}(z_1), f_{n_3}^{(1)}(z_1), \ldots,$$

that is, $\{f_{n_k}^{(1)}\}$ converges at z_1. At the point z_2, as the sequence $\{f_{n_k}^{(1)}(z_2)\}$ of complex numbers is bounded, we can extract a convergent subsequence

$$f_{n_1}^{(2)}(z_2), f_{n_2}^{(2)}(z_2), f_{n_3}^{(2)}(z_2), \ldots,$$

so that $\{f_{n_k}^{(2)}\}$ converges at z_2 and z_1. Continuing in this fashion yields subsequences $\{f_{n_k}^{(p)}\}$ which converge at $z_1, z_2, \ldots, z_p$, for each $p \in \mathbb{N}$. Again taking the diagonal sequence $\{f_{n_k}^{(k)}\}$ we find that this sequence converges at every z_n.

We proceed to show that the diagonal sequence converges uniformly on compact subsets. Let us rename the sequence $\{g_k\} = \{f_{n_k}^{(k)}\}$. Take $K \subseteq \Omega$ to be compact and $\varepsilon > 0$. As $\mathcal{F}$ is equicontinuous on K (Theorem 1.6.3), there exists $\delta = \delta(\varepsilon, K) > 0$ such that

$$|g_n(z) - g_n(z')| < \frac{\varepsilon}{3}, \qquad n = 1, 2, 3, \ldots$$

whenever $z, z' \in K$, $|z - z'| < \delta$. The compactness of K also implies that $K \subseteq \cup_{k=1}^{k_0} D(z_k; \delta)$, after possibly renaming the points $z_1, \ldots, z_{k_0}$. Then there is an integer n_0 such that $n, m \geq n_0$ implies

$$|g_n(z_k) - g_m(z_k)| < \frac{\varepsilon}{3},$$

for $k = 1, \ldots, k_o$.

Finally, for any $z \in K$, $z \in D(z_i; \delta)$ for some $1 \leq i \leq k_0$, and

$$
\begin{aligned}
|g_n(z) - g_m(z)| &\leq |g_n(z) - g_n(z_i)| + |g_n(z_i) - g_m(z_i)| + |g_m(z_i) - g_m(z)| \\
&< \frac{\varepsilon}{3} + \frac{\varepsilon}{3} + \frac{\varepsilon}{3} = \varepsilon.
\end{aligned}
$$

We conclude that $\{g_n\}$ converges uniformly on K to an analytic function (by the Weierstrass Theorem (§1.4)), and the proof is complete.

The theorem was established independently by Koebe [1908]. With suitable modifications to the proof, a Riemann surface version ensues (cf. Ford [1951], p. 268).

Remark 1. The underlying domain Ω in Montel's theorem can also be permitted to lie in $\widehat{\mathbb{C}}$, the extended complex plane, which is desirable for instance, in the study of discrete groups (§5.5). In fact, the functions $g(z) = f\left(\frac{1}{z}\right)$ are uniformly bounded in a disk $|z| < \frac{1}{R}$ for some sufficiently large R, and thus constitute a normal family there. Then $\mathcal{F}$ is normal at $z = \infty$ and hence in all of Ω.

Remark 2. Note that the theorem is actually valid in any *open set* in $\widehat{\mathbb{C}}$, for its components are only countable in number and the theorem may be applied successively in each. The assertion then follows by taking a diagonal sequence and noting that any compact subset lies in only finitely many components.

It is also possible to establish a partial converse to Montel's theorem (Montel [1910]).

Theorem 2.2.2 *Let $\mathcal{F}$ be a family of analytic functions on a domain Ω such that every sequence of functions in $\mathcal{F}$ has a subsequence converging uniformly on compact subsets to an analytic function. Then $\mathcal{F}$ is locally bounded, and hence equicontinuous on compact subsets of Ω.*

Proof. Suppose that $\mathcal{F}$ is not locally bounded. Then there is some closed disk $\Delta \subseteq \Omega$ such that for each positive integer n, there is a function $f_n \in \mathcal{F}$ and a point $z_n \in \Delta$ with

$$|f_n(z_n)| > n.$$

A subsequence, $\{f_{n_k}\}$, then converges uniformly on Δ to an analytic function f, i.e., for some $k_0 \in \mathbb{N}$ and $k \geq k_0$

$$|f_{n_k}(z) - f(z)| < 1, \qquad z \in \Delta.$$

Thus, if $M = \max_{z \in \Delta} |f(z)| < \infty$,

$$|f_{n_k}(z)| \leq 1 + M, \qquad z \in \Delta,$$

which contradicts the fact that $|f_{n_k}(z_{n_k})| \to \infty$ as $k \to \infty$.

That normality by itself is insufficient for a family of analytic functions to be equicontinuous is illustrated by the following.

Example 2.2.3 Let $f(z)$ be zero-free and analytic in a domain Ω and set $\mathcal{F} = \{cf(z) : c \in \mathbf{R}^+\}$. Then every sequence of functions of $\mathcal{F}$ has a normally convergent subsequence whose limit is either finite or identically infinite. Thus, $\mathcal{F}$ is normal. However, for distinct z, z' with $f(z) \neq f(z')$,

$$|cf(z) - cf(z')| = c|f(z) - f(z')|,$$

and the right-hand side can be made arbitrarily large, so that $\mathcal{F}$ is not equicontinuous or locally bounded.

On the other hand, normality does imply equicontinuity in the context of the spherical metric; see Theorem 3.2.1.

In order to guarantee the hypotheses of Theorem 2.2.2, we can rephrase it in the following way.

Corollary 2.2.4 *Let $\mathcal{F}$ be a normal family of analytic functions on a domain Ω. If for some point $z_0 \in \Omega$, $|f(z_0)| \leq M$, for some constant $M < \infty$, for all $f \in \mathcal{F}$, then $\mathcal{F}$ is locally bounded.*

Corollary 2.2.5 *If $\mathcal{F}$ is a normal family of analytic functions in Ω and $|f(z_0)| \leq M$ for some $z_0 \in \Omega$, all $f \in \mathcal{F}$, then the family $\mathcal{F}' = \{f' : f \in \mathcal{F}\}$ is locally bounded and normal in Ω.*

In fact, since $\mathcal{F}$ is locally bounded, $\mathcal{F}'$ is likewise by Theorem 1.5.2, whence $\mathcal{F}'$ is normal by Montel's theorem.

However, a family of analytic functions can be normal without the corresponding family of derivatives being normal. For example, consider the family $\{nz : n = 1, 2, 3, \ldots\}$ which we have seen is not normal in any neighbourhood of the origin. Define

$$\phi_n(z) = \frac{nz^2}{2} + n, \qquad n = 1, 2, 3, \ldots .$$

Then for all $z \in U$,

$$|\phi_n(z)| \geq n - \left|\frac{nz^2}{2}\right| \geq \frac{n}{2},$$

and so $\{\phi_n\}$ is normal in U, yet $\{\phi_n'\}$ is not.

One can sometimes establish that, for a given family of analytic functions, the corresponding family of derivatives is locally bounded. This, in turn, is sufficient for the normality of the original family (Montel [1907]).

Theorem 2.2.6 *If $\mathcal{F}$ is a family of analytic functions on Ω and the family $\mathcal{F}' = \{f' : f \in \mathcal{F}\}$ is locally bounded, then $\mathcal{F}$ is normal in Ω.*

Proof. Given a sequence $\{f_n\} \subseteq \mathcal{F}$, the corresponding sequence of derivatives $\{f_n'\} \subset \mathcal{F}'$ has a convergent subsequence $\{f_{n_k}'\}$ converging normally to an analytic function $g(z)$ in Ω. First let $\Delta \subseteq \Omega$ be a closed disk, and fix $z_0 \in \Delta$. Then, integrating over a straight-line path

$$f_{n_k}(z) = f_{n_k}(z_0) + \int_{z_0}^z f_{n_k}'(\zeta)\, d\zeta, \qquad z \in \Delta.$$

(i) If $\sup_k |f_{n_k}(z_0)| < \infty$, then there is a convergent subsequence $\{f_p\} \subseteq \{f_{n_k}\}$ such that

$$\lim_{p \to \infty} f_p(z) = \lim_{p \to \infty} f_p(z_0) + \int_{z_0}^z g(\zeta)\, d\zeta = F(z)$$

uniformly in Δ, where $F(z)$ is analytic. Note that $F' = g$ in Δ.

(ii) If $\sup_k |f_{n_k}(z_0)| = \infty$, then a subsequence $\{f_p\} \subseteq \{f_{n_k}\}$ converges uniformly to ∞ on Δ.

The proof is completed using a compactness argument. We remark that the theorem is also a trivial consequence of the Marty Theorem (§3.3).

As an illustration we have:

Example 2.2.7 Given an analytic function $f = u + iv$ in U, the *Dirichlet integral* of f is defined as

$$D_U(f) = \int\int_U |f'(z)|^2\, dx\, dy = \int\int_U |\operatorname{grad} u|^2\, dx\, dy = D_U(u).$$

The Dirichlet integral of f measures the planar area of the image $f(U)$ with due regard for multiplicity, that is to say, the area of the Riemann surface onto which $f(U)$ can be mapped bijectively. Then the family

$$\mathcal{F}_M = \{f \text{ analytic in } U : D_U(f) \le M < \infty\}$$

is normal. To see this, apply the areal mean value property (cf. Radó [1971], p. 23) to $\mathcal{R}e\,(f')^2$, $\mathcal{I}m(f')^2$, giving, for $z \in U$,

$$[f'(z)]^2 = \frac{1}{\pi\rho^2} \int_0^\rho \int_0^{2\pi} [f'(z + re^{i\theta})]^2 \, r \, dr \, d\theta,$$

where $\rho = 1 - |z|$. Then, setting $\zeta = \xi + i\eta$,

$$|f'(z)|^2 \le \frac{1}{\pi(1 - |z|)^2} \int\int_U |f'(\zeta)|^2 \, d\xi \, d\eta \le \frac{M}{\pi(1 - |z|)^2}, \qquad z \in U,$$

and we conclude that $\mathcal{F}' = \{f' : f \in \mathcal{F}_M\}$ is locally bounded in U. Hence $\mathcal{F}_M$ is normal by the theorem; cf. also Examples 4.1.2 and 5.4.9.

Actually, the proof demonstrates something further, but at this juncture we simply write the preceding inequality in the form

$$(1 - |z|)|f'(z)| \le \sqrt{\frac{1}{\pi}D_U(f)}, \qquad |z| < 1,$$

and pursue the discussion further in the section on *normal functions* (§5.3, example (vii)). Furthermore, the family

$$\mathcal{G}_M = \{f \text{ analytic in } U : \int\int_U |f(z)|^2 \, dx \, dy \le M < \infty\}$$

is normal by analogous reasoning. Indeed, the power 2 in the preceding two examples may be replaced by $p > 0$ owing to the subharmonicity of $|g|^p$, when g is analytic (cf. Radó [1971], pp. 19 and 23).

A variation of Montel's theorem has been given by Mandelbrojt [1929].

Theorem 2.2.8 *Let $\mathcal{F}$ be a family of zero-free analytic functions in a domain Ω. Then $\mathcal{F}$ is normal in Ω if and only if the corresponding family of functions given by*

$$\tilde{f}(z, w) = f(z)/f(w)$$

is locally bounded on $\Omega \times \Omega$.

Compactness. Given a domain Ω, we define $A(\Omega)$ to be the vector space of analytic functions on Ω. Let $\mathcal{F} \subseteq A(\Omega)$ be locally bounded and closed in the topology of uniform convergence on compact subsets of Ω. Montel's theorem implies $\mathcal{F}$ is sequentially compact, and since $A(\Omega)$ is a complete metric space, $\mathcal{F}$ is compact, that is

Corollary 2.2.9 *Any closed locally bounded subset of $A(\Omega)$ is compact.*

Accordingly, a normal family $\mathcal{F}$ is *compact* if the limit function of every normally convergent subsequence belongs to $\mathcal{F}$. For the compactification of normal families, see Theorem 3.4.1.

2.3   Examples

1. $\mathcal{F} = \{f_n(z) = z^n : n = 1, 2, 3, \ldots\}$ in U. Then $\mathcal{F}$ is uniformly bounded, hence normal in U, but not compact since $0 \notin \mathcal{F}$. In the domain $U' : |z| > 1$, $\{f_n\}$ converges uniformly to ∞ on compact subsets of U', and so $\mathcal{F}$ is normal there as well.

2. $\mathcal{F} = \{f_n(z) = \frac{z}{n} : n = 1, 2, 3, \ldots\}$ is a normal family in $\mathbb{C}$ but not compact.

3. $\mathcal{F} = \{f : f$ analytic in Ω and $|f| \le M\}$. Then $\mathcal{F}$ is normal in Ω and compact.

4. $\mathcal{F} = \{f : f$ analytic in Ω, $\mathcal{R}e\, f > 0\}$. Then the auxiliary family $\mathcal{G} = \{g = e^{-f} : f \in \mathcal{F}\}$ is a uniformly bounded family, hence normal, and it follows that $\mathcal{F}$ is normal, but not compact; cf. also Theorem 5.4.3.

5. $\mathcal{S} = \{f : f$ analytic, univalent in U, $f(0) = 0$, $f'(0) = 1\}$. These are the normalized "schlicht" functions in U. A well-known growth condition for $f \in \mathcal{S}$ asserts (cf. Pommerenke [1975], p. 21)

$$|f(z)| \le \frac{|z|}{(1 - |z|)^2}, \qquad z \in U.$$

Hence $\mathcal{S}$ is locally bounded and thus a normal family. Moreover, if $f_n \to f$ uniformly on compact subsets of U, then f is analytic in U and either univalent or constant, by Corollary 1.4.1. As $f_n'(0) \to f'(0)$, we have $f'(0) = 1$ so that f is univalent in U. Clearly $f(0) = 0$, and $f \in \mathcal{S}$, i.e., $\mathcal{S}$ is compact.

Remark. There is a complementary class of univalent functions which admit the representation

$$g(z) = z + \frac{b_1}{z} + \frac{b_2}{z^2} + \ldots, \qquad |z| > 1.$$

This class is also normal and compact and is discussed in §5.1 where it arises naturally in an application of conformal mappings.

6. $\mathcal{F} = \left\{ t(z) = e^{i\gamma} \frac{z+a}{1+\bar{a}z} : \gamma \text{ real}, |a| < 1 \right\}$ is a normal family in U since $|t(z)| < 1$, $z \in U$, and $\mathcal{F}$ becomes compact if $|a| \leq r_0 < 1$.

7. $\mathcal{N}_M = \{f : f \text{ analytic in } U, \ m(R, f) \leq M < \infty, \ 0 < R < 1\}$. This is a subfamily of the functions of *bounded type*. From §1.8, $m(R, f) = \frac{1}{2\pi} \int_0^{2\pi} \log^+ |f(Re^{i\theta})| \, d\theta$, and by Theorem 1.8.2

$$\log M(r, f) \leq \frac{R+r}{R-r} m(R, f) \leq \frac{R+r}{R-r} M, \qquad 0 \leq r < R < 1.$$

This means that $\mathcal{N}_M$ is locally bounded in U, hence normal, and indeed compact. Refer to Theorem 3.6.9 for the meromorphic counterpart.

8. $\mathcal{F} = \{f_\alpha \text{ analytic in } U : f_\alpha(z) = \sum_{n=1}^{\infty} c_n^{(\alpha)} z^n, \ |c_n^{(\alpha)}| \leq M\}$. Then

$$|f_\alpha(z)| \leq \frac{M}{1-r}$$

for $|z| \leq r < 1$, implying $\mathcal{F}$ is normal in U. $\mathcal{F}$ is easily seen to be compact (cf. the discussion preceding Example 5.1.2).

9. $\mathcal{F} = \{f \text{ analytic in } \Omega : |f(z) - a| > m\}$ for some fixed $a \in \mathbb{C}$, $m > 0$. Then the family $\mathcal{G} = \left\{ g(z) = \frac{1}{f(z)-a} : f \in \mathcal{F} \right\}$ satisfies $|g(z)| < \frac{1}{m}$ in Ω, and is therefore normal. We claim $\mathcal{F}$ is also normal in Ω. In fact, let $\{f_n\} \subseteq \mathcal{F}$ and consider the corresponding sequence $\{g_n\} \subseteq \mathcal{G}$; it has a subsequence $\{g_{n_k}\}$ which converges normally to an analytic function $g(z)$. As no g_{n_k} vanishes in Ω, either $g(z)$ does not vanish in Ω or $g \equiv 0$ by Hurwitz's theorem.

Suppose that $g(z) \neq 0$ in Ω. Then corresponding to g_{n_k}, for $k = 1, 2, 3, \ldots$, are functions $f_{n_k} \in \mathcal{F}$ as well as

$$f(z) = a + \frac{1}{g(z)}.$$

Therefore

$$f(z) - f_{n_k}(z) = \frac{1}{g(z)} - \frac{1}{g_{n_k}(z)} = \frac{g_{n_k}(z) - g(z)}{g(z) g_{n_k}(z)}.$$

If $K \subseteq \Omega$ is compact, then $|g(z)| > \alpha > 0$ for $z \in K$. On the other hand, taking k sufficiently large, given $0 < \varepsilon < \alpha$,

$$|g_{n_k}(z) - g(z)| < \varepsilon, \qquad z \in K.$$

As a consequence

$$|g_{n_k}(z)| > \alpha - \varepsilon,$$

and

$$|f(z) - f_{n_k}(z)| < \frac{\varepsilon}{\alpha(\alpha - \varepsilon)},$$

for k sufficiently large, $z \in K$.

When $g \equiv 0$, we have $|g_{n_k}(z)| < \varepsilon$ for k sufficiently large, $z \in K$, so that

$$|f_{n_k}(z)| = \left| \frac{1}{g_{n_k}(z)} + a \right| > \frac{1}{\varepsilon} - |a|.$$

As the right-hand side can be made arbitrarily large, $\{f_n\}$ converges uniformly to ∞ on K.

As this example shows, $\mathcal{F}$ is normal if it omits some open subset of the plane. It will be shown in §2.7 (Fundamental Normality Test) that the omitted set can, in fact, consist of merely two points.

10. If $\mathcal{S}$ is the class of schlicht functions of Example 5, then by Corollary 2.2.5 the family of derivatives, $\mathcal{S}'$ is normal in U. Furthermore, the family $\mathcal{S}'_{-1} = \left\{ \frac{1}{f'} : f \in \mathcal{S} \right\}$ is locally bounded and hence normal. For, suppose to the contrary that there is some compact set $\Delta \subseteq U$ for which

$$|f'_n(z_n)| < \frac{1}{n}, \qquad z_n \in \Delta, \ f_n \in \mathcal{S}, \ n = 1, 2, 3, \ldots.$$

By the compactness of Δ and $\mathcal{S}$, we may assume $z_n \to z_0$, $f'_n \to f'$ uniformly on Δ, where f is univalent. However, this contradicts

$$f'(z_0) = \lim_{n \to \infty} f'_n(z_n) = 0,$$

proving that $\mathcal{S}'_{-1}$ is locally bounded.

We shall require an extension of the results of Examples 5 and 10 to an arbitrary domain Ω. Consider the family

$$\mathcal{F} = \{f \text{ analytic univalent in } \Omega : |f(a_0)| \le A, \ B \le |f'(a_0)| \le C\},$$

where $a_0 \in \Omega$, and A, B, C are positive constants. Our aim is to prove that: $\mathcal{F}$, as well as $\mathcal{F}' = \{f' : f \in \mathcal{F}\}$ and $\mathcal{F}'_{-1} = \left\{ \frac{1}{f'} : f \in \mathcal{F} \right\}$ are locally bounded (hence normal) in Ω. To this end, choose an arbitrary point $a \in \Omega$, an open disk $D(a; r) \subseteq \Omega$, and define

$$\tilde{f}(\zeta) = \frac{f(a + r\zeta) - f(a)}{r f'(a)}, \qquad \zeta \in U, \ f \in \mathcal{F}.$$

Then each $\tilde{f}$ belongs to $\mathcal{S}$ and are thus locally bounded, giving

$$|\tilde{f}(\zeta)| \le m,$$

for all $\zeta \in K(0; r/2)$, say, $f \in \mathcal{F}$. Hence for each $f \in \mathcal{F}$,

$$|f(z)| \leq mr|f'(a)| + |f(a)|,$$

for all $z \in K(a; r/2)$.

Suppose that now the following condition holds:

$$M_a = \sup\{|f(a)| + |f'(a)| : f \in \mathcal{F}\} < \infty. \qquad (2.1)$$

Then

$$|f(z)| \leq mrM_a + M_a = C < \infty,$$

for all $z \in K(a; r/2)$, $f \in \mathcal{F}$. Furthermore, as in Theorem 1.5.2,

$$|f'(z)| \leq \frac{8C}{r} = C',$$

for all $z \in D(a; r/4)$, $f \in \mathcal{F}$. We have deduced that the validity of condition (2.1) at a point $a \in \Omega$ implies that both $\mathcal{F}$ and $\mathcal{F}'$ are uniformly bounded in $D(a; r/4)$, given $D(a; r) \subseteq \Omega$.

Verifying that $\mathcal{F}$ and $\mathcal{F}'$ are locally bounded in Ω necessitates a delicate connectedness argument. Indeed, set

$$\Omega_0 = \{z \in \Omega : \mathcal{F}, \mathcal{F}' \text{ are uniformly bounded in some neighbourhood of } z\}.$$

Clearly $a_0 \in \Omega_0$, and Ω_0 is an open set. To prove that Ω_0', if nonvoid, is also open, let us assume the contrary. That is, suppose there is some point $z' \in \Omega_0'$ about which every open set intersects Ω_0, and take a disk $D(z'; r) \subseteq \Omega$. Then there exists a point $z \in \Omega_0$ with $|z - z'| < \frac{r}{16}$, so that $D(z; r/2) \subseteq \Omega$. Since $\mathcal{F}$ and $\mathcal{F}'$ are uniformly bounded at z, condition (2.1) holds at z, implying $\mathcal{F}$ and $\mathcal{F}'$ are uniformly bounded on $D(z; r/8)$ by the foregoing analysis. Therefore, $\mathcal{F}$ and $\mathcal{F}'$ are uniformly bounded in a neighbourhood of z' and, consequently, $z' \in \Omega_0$, a contradiction. The connectedness of Ω yields $\Omega_0 \equiv \Omega$ and so we obtain the result that $\mathcal{F}$ and $\mathcal{F}'$ are locally bounded and normal in Ω. The local boundedness of $\mathcal{F}'_{-1}$ now follows as in Example 10.

The preceding discussion permits an elementary proof of one version of the

Koebe Distortion Theorem *Let $f(z)$ be analytic univalent in a domain Ω and K a compact subset of Ω. Then there exists a constant $c = c(\Omega, K)$ such that for any $z, w \in K$*

$$\frac{1}{c} \leq \left| \frac{f'(z)}{f'(w)} \right| \leq c.$$

Proof. For each $f \in \mathcal{F} = \{f$ analytic, univalent in $\Omega\}$ and fixed $z_0 \in \Omega$, define

$$g(z) = \frac{f(z) - f(z_0)}{f'(z_0)}, \qquad z \in \Omega.$$

Then $g(z_0) = 0$, $g'(z_0) = 1$, and $g(z)$ is analytic univalent in Ω. Thus, $\{g' : f \in \mathcal{F}\}$ and $\left\{\frac{1}{g'} : f \in \mathcal{F}\right\}$ are locally bounded, implying

$$\frac{1}{c} \leq |g'(z)| \leq c, \qquad z \in K,$$

that is,

$$\frac{1}{c} \leq \left|\frac{f'(z)}{f'(z_0)}\right| \leq c.$$

Multiplying through with another application of this inequality gives

$$\frac{1}{c^2} \leq \left|\frac{f'(z)}{f'(w)}\right| \leq c^2, \qquad z, w \in K.$$

This result was used by Koebe in this theory of uniformization.

2.4   Vitali-Porter Theorem

As we have seen in Theorem 1.6.4, a locally bounded sequence of analytic functions converging pointwise converges uniformly on compact sets. A much stronger result in this vein was proved independently by Vitali [1903, 1904] and Porter [1904–05].

Vitali-Porter Theorem *Let $\{f_n\}$ be a locally bounded sequence of analytic functions in a domain Ω such that $\lim_{n\to\infty} f_n(z)$ exists for each z belonging to a set $E \subseteq \Omega$ which has an accumulation point in Ω. Then $\{f_n\}$ converges uniformly on compact subsets of Ω to an analytic function.*

Proof. As $\{f_n\}$ is normal, extract a subsequence $\{f_{n_k}\}$ which converges normally to an analytic function f. Then $\lim_{k\to\infty} f_{n_k}(z) = f(z)$ for each $z \in E$.

Suppose, however, that $\{f_n\}$ does not converge uniformly on compact subsets of Ω to f. Then there exists some $\varepsilon > 0$, a compact subset $K \subseteq \Omega$, as well as a subsequence $\{f_{m_j}\}$ and points $z_j \in K$ satisfying

$$|f_{m_j}(z_j) - f(z_j)| \geq \varepsilon, \qquad j = 1, 2, 3, \ldots . \tag{2.2}$$

Now $\{f_{m_j}\}$ itself has a subsequence which converges uniformly on compact subsets to an analytic function g, and $g \not\equiv f$ in view of (2.2). However, since f and g must agree at all points of E, the Identity Theorem for analytic functions (cf. Ahlfors [1979], p. 127) implies $f \equiv g$ on Ω, a contradiction which establishes the theorem.

There are *ab initio* proofs in Lindelöf [1913] and Jentzsch [1918].

The Vitali-Porter Theorem was reformulated in terms of meromorphic functions by Montel (Theorem 3.2.3). Some credit must, of course, be given to a prior result of Stieltjes [1894]: *If $\{f_n\}$ is a locally bounded sequence of analytic functions in Ω and $\{f_n\}$ converges uniformly on any subdomain of Ω, then $\{f_n\}$ converges uniformly on every compact subset to an analytic function.* For an interesting variation on this theme, cf. Milloux [1948].

The following application is essentially due to Montel [1910], pp. 27–28.

Theorem 2.4.1 *Let Ω be a domain and suppose that $f : [0,1] \times \Omega \to \mathbb{C}$ satisfies*

(i) $|f(t,z)| \leq M$ *for all* $t \in [0,1], \cdot z \in \Omega$;

(ii) $f(t,z)$ *is a continuous function of* t *for each fixed* $z \in \Omega$;

(iii) $f(t,z)$ *is an analytic function of* z *for each fixed* $t \in [0,1]$.

If $g(t)$ is a function of bounded variation on $[0,1]$, then the function defined by

$$F(z) = \int_0^1 f(t,z)\,dg(t),$$

is analytic in Ω.

Proof. From the definition of the Riemann-Stieltjes integral we can write

$$F_n(z) = \sum_{k=1}^n f\big(t_k^{(n)}, z\big)\Big[g\big(t_k^{(n)}\big) - g\big(t_{k-1}^{(n)}\big)\Big],$$

where $F_n(z) \to F(z)$ at each $z \in \Omega$, and $\{t_k^{(n)}\}$ is a set of partitions of $[0,1]$. Then $F_n(z)$ is analytic in Ω and satisfies $|F_n(z)| \leq M \cdot V[g]$, where $V[g]$ is the total variation of g on $[0,1]$. By the Vitali-Porter theorem, $F_n \to F$ uniformly on compact subsets of Ω, establishing the analyticity of $F(z)$.

Montel's theorem (§2.2) can also be deduced from the Vitali-Porter Theorem; *thus, the two results are actually equivalent.* In fact, let $\mathcal{F}$ be a locally bounded family of analytic functions on Ω, choose $\{f_n\} \subseteq \mathcal{F}$, and let E be a countable subset of Ω having an accumulation point in Ω; say $E = \big\{z_n = z_0 + \frac{1}{n}$ for all n sufficiently large, $z_0 \in \Omega\big\}$. Since $\{f_n\}$ is bounded at each $z_n \in E$, we can mimic the first part of the proof of the Montel theorem with $E = \{z_n\}$ and apply the Cantor diagonal process. This yields a subsequence $\{f_{n_k}\}$ which converges at each $z_n \in E$. The normality of $\mathcal{F}$ then follows from the Vitali-Porter theorem.

A sometimes useful consequence of the Montel and Vitali-Porter theorems is (Montel [1910], p. 21).

Theorem 2.4.2 *Suppose that* $\{f_n\}$ *is a locally bounded sequence in a domain* Ω *such that every convergent subsequence converges to a function* g. *Then* $\{f_n\}$ *converges to* g, *locally uniformly in* Ω.

Proof. Assuming the sequence of complex numbers $\{f_n(\zeta)\}$ does not converge to the complex number $g(\zeta)$, $\zeta \in \Omega$, then there exists $\delta > 0$ and $1 \leq n_1 < n_2 < \ldots$ such that

$$|f_{n_k}(\zeta) - g(\zeta)| > \delta, \qquad k = 1, 2, 3, \ldots.$$

Montel's theorem implies $\{f_{n_k}\}$ has a convergent subsequence. However, its limit cannot be g, violating the hypotheses, and so $f_n(z) \to g(z)$ for all $z \in \Omega$. The Vitali-Porter Theorem then provides the locally uniform convergence.

In the same *genre* as the Vitali-Porter Theorem, and a consequence of it (although not historically) is the

Osgood's Theorem [1901–02] *If* $\{f_n\}$ *is a sequence of analytic functions in a domain* Ω *such that* $\lim_{n\to\infty} f_n(z) = f(z) \in \mathbb{C}$ *for each* $z \in \Omega$, *then* $\{f_n\}$ *converges locally uniformly on a dense subset* $\Omega_0 \subset \Omega$. *In particular,* $f(z)$ *is analytic in* Ω_0.

Proof. For each positive integer N, define

$$\Omega_N = \{z \in \Omega : |f_n(z)| \leq N, \ n = 1, 2, 3, \ldots\}.$$

Then each Ω_N is a closed subset of Ω, and $\Omega = \cup_{N=1}^{\infty} \Omega_N$ since $\lim_{n\to\infty} f_n(z)$ exists. Therefore, every open set O with $\overline{O} \subseteq \Omega$ satisfies

$$\overline{O} = \bigcup_{N=1}^{\infty} \overline{O} \cap \Omega_N.$$

According to the Baire Category Theorem (cf. Hewitt and Stromberg [1965], p. 68), there is some $\overline{O} \cap \Omega_{N_0}$, $N_0 = N_0(O)$, which is not nowhere dense, that is, contains an open disk $\Delta(O) \subseteq \overline{O}$, with $|f_n(z)| \leq N_0$ for all n, and $z \in \Delta(O)$. Setting

$$\Omega_0 = \cup\{\Delta(O) : \overline{O} \subseteq \Omega\},$$

we find that Ω_0 is an open and dense subset of Ω. Since any compact subset $K \subseteq \Omega_0$ is covered by finitely many of the disks $\Delta(O)$, the family $\{f_n\}$ is uniformly bounded on the union of these disks. The conclusion of the theorem is then a consequence of the Vitali-Porter Theorem.

Another proof can be found in Montel [1932]. For the meromorphic counterpart, see Carathéodory [1958, Vol. I], pp. 190–191.

Generalizations. A considerable literature has built up dealing with generalizations of the Vitali-Porter Theorem. One such, due to Carathéodory and Landau [1911], relaxing the boundedness conditions, was the likely inspiration for Montel's Fundamental Normality Test (§2.7). Others, principally due to Montel [1907, 1917, 1927], Blaschke [1915], Khintchine [1922–24, 1923], Nevanlinna and Nevanlinna [1922], Ostrowski [1922–23], F. Riesz [1922–23], Kunugui [1942], relax the condition on the limit point belonging to Ω, or replace the interior sequence of points entirely by a portion of $\partial\Omega$. We consider two results in this vein.

It is well-known that if $f(z)$ is a bounded analytic function, $f \not\equiv 0$, in the unit disk U, then the zeros of $f(z)$, $\{a_n\}$ satisfy $\sum_n (1 - |a_n|) < \infty$ (cf. Rudin [1974], Theorem 15.23). Using this fact, Blaschke [1915] established the following analogue of the Vitali-Porter Theorem.

Theorem 2.4.3 *If $\{f_n\}$ is a uniformly bounded sequence of analytic functions in U such that $\{f_n\}$ converges at each $a_n \in U$, $n = 1, 2, 3, \ldots$, and $\sum_{n=1}^{\infty}(1 - |a_n|) = \infty$, then $\{f_n\}$ converges uniformly on compact subsets of U to an analytic function.*

Proof. As $\{f_n\}$ is normal and $|f_n| \leq M$, say, there is at least one limit function $f(z)$, and $|f| \leq M$ in U. If the whole sequence $\{f_n\}$ does not converge normally to f, there must be another limit function g, with $f(a) \neq g(a)$ for some $a \in U$. But f and g are bounded in U with $f(a_n) - g(a_n) = 0$, which, by the preceding remarks, implies $f \equiv g$, a contradiction, and the result is proved.

By virtue of Theorem 1.8.2, the uniform boundedness of $\{f_n\}$ may be replaced by the *Nevanlinna condition*

$$m(r, f_n) = \int_0^{2\pi} \log^+ |f_n(re^{i\theta})|\, d\theta \leq M < \infty, \qquad 0 \leq r < 1,$$

for $n = 1, 2, 3, \ldots$ (Nevanlinna and Nevanlinna [1922]), or indeed by $m(r, f_n) < \phi(r)$, where ϕ is an increasing function of r, as $r \to 1$.

2.5   Zeros of Normal Families

For a normal family of analytic functions, the number of zeros the family has, in general, turns out to be locally bounded. More precisely

Theorem 2.5.1 *Let $\mathcal{F}$ be a normal family of analytic functions in the domain Ω, and suppose that each limit function is not equal to a constant a. Then for every compact subset K of Ω, there exists a constant $M = M(K)$, such that the number of zeros of $f(z) - a$ in K does not exceed M, for each $f \in \mathcal{F}$.*

Proof. Assume K is a compact subset of Ω for which the number of zeros of $f(z) - a$ has no bound, for $f \in \mathcal{F}$. Then there is a function $f_1 \in \mathcal{F}$

such that $f_1(z) - a$ has at least one zero in K, a function $f_2 \in \mathcal{F}$ such that $f_2(z) - a$ has at least two zeros in K, and thus a sequence $\{f_n\} \subseteq \mathcal{F}$ such that $f_n(z) - a$ has at least n zeros in K. It follows that there is a subsequence $\{f_{n_k}\}$ that converges uniformly on K to a function f.

If $f(z) \equiv \infty$, then for k sufficiently large, $|f_{n_k}(z)| > |a|$, $z \in K$, contradicting the fact that $f_{n_k}(z) - a$ has at least n_k zeros in K. Hence $f(z)$ is analytic in Ω. Since $f(z) \not\equiv a$, the zeros of $f(z) - a$ in K are isolated, say at the points $z_1, z_2, \ldots, z_\ell$ with multiplicities $\alpha_1, \alpha_2, \ldots, \alpha_\ell$, respectively. Consider a set of mutually disjoint disks $D(z_i; r_i)$, which may be assumed to lie in K. For k sufficiently large, $f_{n_k}(z) - a$ has the same number of zeros in each $D(z_i; r_i)$ as does $f(z) - a$, so that for such k, $f_{n_k}(z) - a$ has $\alpha = \alpha_1 + \ldots + \alpha_\ell$ zeros in $\cup_{i=1}^\ell D(z_i; r_i)$.

In $E = K - \cup_{i=1}^\ell D(z_i; r_i)$, certainly $|f(z) - a| \geq m$, for some constant $m > 0$. Taking $0 < \varepsilon < m$ and k sufficiently large,

$$|f_{n_k}(z) - f(z)| < \varepsilon, \qquad z \in E,$$

and therefore $|f_{n_k}(z) - a| > m - \varepsilon$, $z \in E$. That is, $f_{n_k}(z) - a$ has no further zeros in E, which is incompatible with our assumption that the number of zeros must increase without bound in K. This proves the theorem.

An analogue of this theorem regarding the number of poles of a normal family of meromorphic functions will be given in the next chapter.

2.6   Riemann Mapping Theorem

A very common application of normal families is in the usual proof of the Riemann Mapping Theorem (although a normal families argument is not essential, cf. Burckel [1979], p. 293–303, for the constructive proof of Koebe and Carathéodory). Riemann's original proof was based upon the Dirichlet Principle, a concept which at the time was insufficiently rigorous and was only perfected a half century later by Hilbert (cf. Monna [1975], Chapter IV). The modern approach has evolved from the works of Koebe, Carathéodory, Fejér, and Riesz.

Riemann Mapping Theorem   *Let Ω be a simply connected domain having at least two boundary points. Then for each $z_0 \in \Omega$, there exists a unique analytic univalent function $w = F(z)$ mapping Ω onto U satisfying the conditions $F(z_0) = 0$, $F'(z_0) > 0$.*

Proof. That the mapping is unique is readily established. For suppose that $F_1(z)$ and $F_2(z)$ are as in the theorem. Then the function

$$L(z) = F_1\big(F_2^{-1}(w)\big)$$

is an analytic univalent mapping of U onto itself. By the Schwarz Lemma (§1.4), $|L(w)| \leq |w|$, implying $|F_1(z)| \leq |F_2(z)|$. Reversing the roles of F_1

and F_2 gives $|F_1(z)| \equiv |F_2(z)|$ in Ω. Then $F_1(z)/F_2(z)$ has constant modulus one, and since the quotient is analytic in Ω, we deduce that $F_1 \equiv F_2$ by the Maximum Modulus Theorem and the positivity of their derivatives at z_0.

Next, let us consider the family

$$\mathcal{F} = \{f(z) \text{ analytic univalent in } \Omega : |f(z)| \leq 1, \ f(z_0) = 0, \ f'(z_0) \geq \alpha\},$$

where α is to be determined. We first endeavour to show that $\mathcal{F}$ is non-empty. For if $a \neq b$ are two finite points on $\partial\Omega$, the quotient $\frac{z-a}{z-b}$ is nonzero and finite for $z \in \Omega$. Since Ω is simply connected, it is possible to define a single-valued analytic branch of $\sqrt{\frac{z-a}{z-b}}$. Specifically, by starting with a point $\zeta \in \Omega$, analytic continuation of some branch of the preceding square root throughout Ω yields a single-valued analytic function $g(z)$, by virtue of the Monodromy Theorem (§1.4). Moreover, $g(z)$ is univalent by a simple deduction, and it is also readily verified that $g(z)$ cannot take on both values ω and $-\omega$ for any $\omega \in \mathbb{C}$. Since $g(z)$ maps Ω onto some domain Ω^*, consider $g(z_0) = \omega_0 \in \Omega^*$ and some disk $D(\omega_0; r_0) \subseteq \Omega^*$. It can then be concluded that $D(-\omega_0; r_0)$ lies outside Ω^*, so that

$$|g(z) + \omega_0| \geq r_0, \qquad z \in \Omega,$$

and, consequently, $|\omega_0| \geq \frac{r_0}{2}$, by putting $z = z_0$. Then the function

$$f_0(z) = \frac{r_0}{4} \frac{|g'(z_0)|}{g'(z_0)} \cdot \frac{\omega_0}{|\omega_0|^2} \cdot \frac{g(z) - \omega_0}{g(z) + \omega_0}$$

satisfies, for $z \in \Omega$,

$$|f_0(z)| = \frac{r_0}{4|\omega_0|} \left| \frac{g(z) - \omega_0}{g(z) + \omega_0} \right| = \frac{r_0}{4} \left| \frac{1}{\omega_0} - \frac{2}{g(z) + \omega_0} \right| \leq 1,$$

and $f_0(z_0) = 0$, $f_0'(z_0) = \frac{r_0}{8} \frac{|g'(z_0)|}{|\omega_0|^2} > 0$. As $f_0(z)$ is also analytic univalent in Ω, we have $f_0 \in \mathcal{F}$ and $\mathcal{F} \neq \emptyset$, when we now set $\alpha = \frac{r_0}{8} \frac{|g'(z_0)|}{|\omega_0|^2}$.

We now consider the extremal problem of determining the existence of a function $F \in \mathcal{F}$ which satisfies the condition

$$F'(z_0) = \sup f'(z_0),$$

amongst all $f \in \mathcal{F}$. To this end, let $A = \sup\{f'(z_0) : f \in \mathcal{F}\}$. Then there is a sequence $\{f_n\} \subseteq \mathcal{F}$ for which $f_n'(z_0) \to A$. Since $\mathcal{F}$ is a uniformly bounded family, and hence normal, there is a subsequence $\{f_{n_k}\}$ which converges locally uniformly in Ω to an analytic function F that satisfies $|F(z)| \leq 1$ in Ω, $F(z_0) = 0$. By Cauchy's formula, $f_{n_k}'(z_0) \to F'(z_0)$, implying $0 < \alpha \leq F'(z_0) = A < \infty$. Moreover, $F(z)$ is univalent, in view of Corollary 1.4.1, and so $F \in \mathcal{F}$, thereby solving our extremal problem.

Actually, $\mathcal{F}$ is a compact family, and extremal problems for compact normal families, such as considered above, can be treated in far greater generality, but by similar methods (cf. §5.1).

It is now claimed that the function $F(z)$ is the desired conformal mapping of Ω onto U. Indeed, suppose there is some $w_0 \in U$ with $F(z) \neq w_0$, $z \in \Omega$. As before, a single-valued analytic branch of the function

$$G(z) = \sqrt{\frac{F(z) - w_0}{1 - \overline{w}_0 F(z)}}$$

can be determined in Ω. Clearly $G(z)$ is univalent with $|G(z)| \leq 1$, $z \in \Omega$. Then the function

$$H(z) = \frac{|G'(z_0)|}{G'(z_0)} \cdot \frac{G(z) - G(z_0)}{1 - \overline{G(z_0)}G(z)}$$

is analytic univalent in Ω, $|H(z)| \leq 1$, $H(z_0) = 0$, and since $F \in \mathcal{F}$,

$$H'(z_0) = \frac{|G'(z_0)|}{1 - |G(z_0)|^2} = \frac{1 + |w_0|}{2\sqrt{|w_0|}} F'(z_0).$$

However,

$$1 + |w_0| = 2\sqrt{|w_0|} + (1 - \sqrt{|w_0|})^2 > 2\sqrt{|w_0|}, \qquad |w_0| < 1,$$

and, consequently,

$$H'(z_0) > F'(z_0).$$

This contradicts the extremal property of F' since $H \in \mathcal{F}$, proving the theorem.

Remark 2.6.1 The above Riemann mapping function given by $w = F(z)$, $F(z_0) = 0$, $F'(z_0) > 0$, determines an analytic univalent inverse function

$$z = F^{-1}(w) = f(w) : U \to \Omega,$$

such that $f(0) = z_0$, $f'(0) > 0$. Then

$$z = f(w) = z_0 + a_1 w + a_2 w^2 + \ldots \qquad |w| < 1,$$

with $a_1 > 0$. Taking instead $|w| > 1$ and $z_0 = 0$, we see that

$$z = f\left(\frac{1}{w}\right) = \frac{a_1}{w} + \frac{a_2}{w^2} + \ldots$$

maps $|w| > 1$ onto Ω, sending ∞ to 0.

Suppose now that Ω^* is a simply connected domain in $\widehat{\mathbb{C}}$ with $\infty \in \Omega^*$ and such that $E = \widehat{\mathbb{C}} - \Omega^*$ contains more than one point, i.e., E is a *continuum*. Then the auxiliary transformation

$$z = \frac{1}{W - \alpha}, \qquad \alpha \in E,$$

maps Ω^* onto a simply connected domain $\Omega \subset \mathbb{C}$ sending ∞ to 0. Combining this with $f\left(\frac{1}{w}\right)$ yields

$$W = \frac{1}{z} + \alpha = \frac{1}{f\left(\frac{1}{w}\right)} + \alpha = bw + b_0 + \frac{b_1}{w} + \frac{b_2}{w^2} + \ldots = G(w)$$

which is a univalent (meromorphic) mapping of $|w| > 1$ onto Ω^*, with $b > 0$, and $\infty \to \infty$. Finally,

$$W = H(w) = G\left(\frac{w}{b}\right) = w + c_0 + \frac{c_1}{w} + \frac{c_2}{w^2} + \ldots$$

maps $|w| > b$ onto Ω^*.

Similarly, by mapping Ω^* onto $|w| > b$, and then effecting the transformation (cf. §5.1)

$$\zeta = w + \frac{e^{2i\theta}}{w},$$

one obtains a univalent mapping $\zeta = \phi(W)$ of Ω^* onto the ζ-plane with a slit at inclination θ with respect to the real axis, such that $\infty \to \infty$ and

$$\phi(W) = W + \frac{d_1}{W} + \frac{d_2}{W^2} + \ldots$$

in a neighbourhood of ∞. The representations of $H(w)$ and $\phi(W)$ feature in the proof of Theorem 5.1.5 on *parallel slit mappings*.

We present now an elegant version of the Riemann Mapping Theorem formulated by Hayman [1983], p. 14; [1989], Theorem 9.9, based on an approach of Carathéodory, and invoking the elliptic modular function; cf. Goluzin [1969], p. 255 for a related result of Poincaré, the proof of which, due to Radó [1922–23], also utilizes normality.

We begin with a few preliminaries. Let Ω be a domain (not necessarily simply connected) whose complement contains at least two points in $\mathbb{C}$. Suppose that $w = f(z)$ is analytic in U and assumes values lying in Ω. Then, as usual, we say $f(z)$ maps U *into* Ω. The inverse function will, in general, be many-valued; if it can be analytically continued throughout Ω (without singularities), with values of every branch of the inverse function lying in U, then we say that $w = f(z)$ maps U *onto* Ω. In the event Ω is simply connected, these onto mappings coincide with the one-to-one conformal mappings of U onto Ω by the Monodromy Theorem (§1.4).

Theorem 2.6.2 *Suppose that Ω is a domain whose complement contains at least two points in $\mathbb{C}$. Given any $a_0 \in \Omega$, there exists a unique analytic mapping $F(z)$ (not necessarily injective) of U onto Ω such that $F(0) = a_0$, $F'(0) > 0$.*

Proof. To show uniqueness, assume $F_1(z)$ and $F_2(z)$ are as in the theorem. Then for $z \in U$, the function

$$L(z) = F_2^{-1}\big(F_1(z)\big),$$

initially defined in a sufficiently small neighbourhood of z, for a particular branch of F_2^{-1}, can be continued analytically throughout all of U. The resulting function $L(z)$ is a single-valued analytic mapping (by the Monodromy Theorem) of U onto itself, satisfying $L(0) = 0$, $L'(0) > 0$. It follows that $L(z) = z$ by the Schwarz Lemma, and $F_1(z) \equiv F_2(z)$.

In order to establish the existence, we may suppose that $0, 1 \in \mathbb{C} - \Omega$. We wish to consider the family $\mathcal{F}_\Omega$ of mappings

$$\phi(w) = z$$

of Ω into U which are *univalent*, that is, $\phi(w) = z$ can be indefinitely continued throughout Ω such that all of its branches $\phi(w) = z$ have their values in U and branches at different points are distinct. Thus, if $w_1 \neq w_2$, no branch of $\phi(w_1)$ can be equal to any branch of $\phi(w_2)$. Moreover, for $\phi \in \mathcal{F}_\Omega$, assume

$$\phi(a_0) = 0.$$

We claim that $\mathcal{F}_\Omega$ is nonempty. To see this, consider the modular function (cf. §1.7) $\mu : U \to \mathbb{C} - \{0, 1\}$, and its inverse $\nu(w) = z$, which is univalent, since μ is single-valued. Let S be a Möbius transformation of U onto itself satisfying $S(\nu(a_0)) = 0$, for a given branch of ν. Whence we find that for any real θ,

$$e^{i\theta} S\big(\nu(w)\big) \in \mathcal{F}_\Omega,$$

proving the assertion.

We proceed more or less as in the proof of the Riemann Mapping Theorem. Set

$$0 < A = \sup\{|\phi'(a_0)| : \phi \in \mathcal{F}_\Omega\}, \tag{2.3}$$

and let $D(a_0; R)$ be the largest disk about a_0 which lies wholly in Ω. Then for each $\phi \in \mathcal{F}_\Omega$, the branches are single-valued analytic in $D(a; R)$. Choose $\{\phi_n\} \subseteq \mathcal{F}_\Omega$ such that $|\phi_n'(a_0)| \to A$. As $\mathcal{F}_\Omega$ is a bounded family, the aggregate of branches of the ϕ_n's has a normally convergent subsequence in $D(a_0; R)$ by Montel's theorem, so we shall assume that $\phi_n(w) \to \Phi(w)$, locally uniformly in $D(a_0; R)$. Moreover, $\Phi(a_0) = 0$, and $|\Phi'(a_0)| = A$.

We must now show that the convergence of $\phi_n(w) \to \Phi(w)$ is uniform in a neighbourhood of any path γ in Ω. Let γ be parametrized by $w = \alpha(t)$, $0 \leq t \leq 1$. For $0 = t_0 < t_1 < \ldots < t_m = 1$, set $w_i = \alpha(t_i)$, and ensure that for some $r > 0$, the closed disks $K_i = K(w_i; r)$ all lie in Ω and include the arc $w_i w_{i+1}$, $i = 0, \ldots, m-1$. Suppose that $\phi_n(w)$ converges in K_j, but does not converge at some point ω in K_{j+1}, taking the branch of $\phi_n(w)$ in K_{j+1} obtained from that in K_j by direct analytic continuation. Then it is possible to find two subsequences, say $\phi_{n_1}(\omega)$ and $\phi_{n_2}(\omega)$, which converge to different values. In view of the uniform boundedness of all the ϕ's in a neighbourhood of $K_j \cup K_{j+1}$, we can find two further subsequences of $\phi_{n_1}(w)$ and $\phi_{n_2}(w)$ that converge uniformly in a neighbourhood $N \supseteq K_j \cup K_{j+1}$ to analytic functions $\psi_1(w)$ and $\psi_2(w)$, respectively. But $\phi_n(w)$ converges in K_j, implying $\psi_1(w) = \psi_2(w)$ in K_j, and hence in K_{j+1}, contradicting $\psi_1(\omega) \neq \psi_2(\omega)$. We conclude that $\phi_n(w)$ converges in K_{j+1}, as desired. Therefore , all the analytic continuations of the $\phi_n(z)$ also converge to a limit $\Phi(z)$ in Ω with $|\Phi(z)| < 1$ there.

In order to show univalence, suppose that a branch of $\Phi(w_1)$ coincides with a branch of $\Phi(w_2)$, and let this common value be denoted by ζ. The function $\Phi(w) - \zeta$ cannot vanish identically, so that by the Hurwitz Theorem (§1.4), $\phi_n(w) - \zeta$ must have a zero in a neighbourhood of w_1 and of w_2, for all n sufficiently large. But his contradicts the univalence of these functions ϕ_n, thus establishing the univalence of Φ, and $\Phi \in \mathcal{F}_\Omega$. We next show that $\Phi(w)$ assumes every value in U, since then the inverse $w = \Phi^{-1}(z)$ is single-valued analytic and maps U onto Ω. For an appropriately chosen θ, so that $e^{i\theta}\Phi'(a_0) > 0$, then $e^{-i\theta}\Phi^{-1}(z)$ yields the desired mapping of the theorem.

Let us assume that $\Phi(w) \neq z_0$ for some $z_0 \in U$. Define mappings F and G by

$$F(w) = \left(\frac{\Phi(w) - z_0}{1 - \overline{z}_0 \Phi(w)} \right)^{\frac{1}{2}}, \quad G(w) = \left(\frac{F(w) - F(a_0)}{1 - \overline{F(a_0)}F(w)} \right).$$

Since $\Phi(w) \neq z_0$, $F(z)$ and $G(z)$ are analytic mappings of Ω into U, with $G(a_0) = 0$. Moreover, $G(w_1) = G(w_2)$ implies $F(w_1) = F(w_2)$, which in turn, yields $\Phi(w_1) = \Phi(w_2)$, and so $w_1 = w_2$. As a consequence, $G \in \mathcal{F}_\Omega$. However, by the chain rule

$$G'(a_0) = \left[\frac{dG}{dF} \frac{dF}{d\Phi} \frac{d\Phi}{dw} \right]_{w=a_0} = \frac{1 + |z_0|}{2|z_0|^{\frac{1}{2}}} \Phi'(a_0),$$

and so $|G'(a_0)| > |\Phi'(a_0)| = A$. This contradicts (2.3) and proves that $\Phi(w)$ assumes all values $z \in U$, thus establishing the theorem.

Other applications of the normal family method to conformal mappings can be found in the researches of de Possel [1931] on the conformal mappings of a multiply connected domain onto a parallel slit domain (cf. §5.1), of Biernacki [1936] on the comparison of two conformal representations of the unit disk, one *subordinate* to the other, and of J. Ferrand [1942] on a study of conformal mappings in a neighbourhood of the boundary, to name but a few.

2.7 Fundamental Normality Test

In 1912 Montel presented his *Critère fondamental* for a family of analytic functions to be normal. The idea for the hypothesis of the theorem is implicit in the paper of Carathéodory and Landau [1911], who established the Vitali-Porter Theorem by replacing the requirement of boundedness by the condition that each function omits two fixed values. The Fundamental Normality Test (FNT) has various far-reaching consequences which are discussed in subsequent sections. As the FNT lies at the very core of the subject of normal families, it is presented in the text from five differing viewpoints, the first one here being essentially the original proof of Montel based on the modular function.

Fundamental Normality Test *Let $\mathcal{F}$ be the family of analytic functions on a domain Ω which omit two fixed values a and b in $\mathbb{C}$. Then $\mathcal{F}$ is normal in Ω.*

Proof. One may assume that the exceptional values a and b are 0 and 1, respectively, by considering the family

$$\widehat{\mathcal{F}} = \left\{ \widehat{f}(z) = \frac{f(z) - a}{b - a} : f \in \mathcal{F} \right\},$$

which does omit 0 and 1, and whose normality is equivalent to that of $\mathcal{F}$. Moreover, since normality is a local property, we may additionally assume Ω is the open disk $|z| < 1$.

Let $\mu(\zeta) = w$ be the elliptic modular function, and let $z_0 \in \Omega$ be given. For any $f \in \mathcal{F}$, choose the single-valued analytic branch of $\nu(w) = \mu^{-1}(w)$ for which $\nu\big(f(z_0)\big)$ is the principal value. Starting with this initial value, it is possible to uniquely define $\nu\big(f(z)\big)$ in a sufficiently small neighbourhood of z_0. Then $\nu\big(f(z)\big)$ may be continued analytically throughout (the disk) Ω since each $f(z)$ is analytic in Ω and takes on values only lying in the domain of $\nu(w)$. This results in a single-valued analytic function being defined by the Monodromy Theorem (§1.4), $\tilde{f}(z) = \nu\big(f(z)\big)$, that satisfies $|\tilde{f}(z)| < 1$ for $z \in \Omega$, $f \in \mathcal{F}$.

Let $\{f_n\}$ be an arbitrary sequence in $\mathcal{F}$, and let $\alpha \in \widehat{\mathbb{C}}$ be an accumulation point of the sequence $\{f_n(z_0)\}$. We consider four cases.

Case (i) : $\alpha \neq 0, 1, \infty$.

Let $\{f_{n'}\} \subseteq \{f_n\}$ be a subsequence such that $\{f_{n'}(z_0)\}$ converges to α as $n' \to \infty$. Then the sequence $\{\tilde{f}_{n'}\}$ itself has a subsequence $\{\tilde{f}_{n_k}\}$ which converges uniformly on compact subsets of Ω to an analytic function $F(z)$ by Montel's theorem, and $|F(z)| \leq 1$. If equality held for any $z \in \Omega$, then $F(z) \equiv e^{i\theta}$ for some real θ, implying $|\tilde{f}_{n_k}(z)| \to 1$ as $k \to \infty$, i.e., $|\nu\big(f_{n_k}(z)\big)| \to 1$. But this happens only if $f_{n_k}(z) \to 0, 1$, or ∞. We conclude that $|F(z)| < 1$ in Ω.

Now take an arbitrary compact subset $K \subseteq \Omega$. Then for some constant m, $|F(z)| \leq m < 1$ on K, and by the uniform convergence of $\tilde{f}_{n_k}$ to F on K there exists m' such that $|\tilde{f}_{n_k}(z)| \leq m' < 1$, $z \in K$, for all k sufficiently large. As the modular function $\mu(\zeta)$ is analytic in U, it is bounded in $|\zeta| \leq m'$, say $|\mu(\zeta)| \leq M$. Then for all k sufficiently large and $z \in K$,

$$|f_{n_k}(z)| = |\mu\left(\nu\left(f_{n_k}(z)\right)\right)| \leq M,$$

i.e., $\{f_{n_k}\}$ is uniformly bounded on compact subsets of Ω. Consequently, this sequence in turn has a subsequence which converges uniformly on compact subsets of Ω to an analytic function.

Case (ii) : $\alpha = 1$.

Let $\{f_{n'}\} \subseteq \{f_n\}$ be a subsequence such that $f_{n'}(z_0) \to 1$ as $n' \to \infty$, and define $g_{n'}(z) = \sqrt{f'_n(z)}$ to be a branch for which $\lim_{n' \to \infty} g_{n'}(z_0) = -1$. Then $g_{n'}$, $n' = 1, 2, 3, \ldots$, are analytic in Ω, omit the values 0 and 1, with limit point at z_0 being -1. By Case (i), there is a subsequence $\{g_{n_k}\} \subseteq \{g_{n'}\}$ which converges uniformly on compact subsets of Ω to an analytic function, and so does $\{g_{n_k}^2\} = \{f_{n_k}\}$.

Case (iii) : $\alpha = 0$.

Again choose a subsequence $\{f_{n'}\} \subseteq \{f_n\}$ such that $f_{n'}(z_0) \to 0$ as $n' \to \infty$, and define $g_{n'}(z) = 1 - f_{n'}(z)$, $n' = 1, 2, 3, \ldots$. Then the functions $g_{n'}(z)$ are analytic in Ω, omit the values 0 and 1, with $\lim_{n' \to \infty} g_{n'}(z_0) = 1$. Thus, $\{g_{n'}\}$ comes under Case (ii), and there exists a subsequence which converges uniformly on compact subsets of Ω to an analytic function, and likewise for $\{f_{n'}\}$.

Case (iv) : $\alpha = \infty$.

Here there is a subsequence $\{f_{n'}\} \subseteq \{f_n\}$ for which $f_{n'}(z_0) \to \infty$ as $n' \to \infty$. Then the functions $g_{n'}(z) = 1/f_{n'}(z)$, $n' = 1, 2, 3, \ldots$, are analytic in Ω, omit 0 and 1, with $\lim_{n' \to \infty} g_{n'}(z_0) = 0$. Thus, by Case (iii) there is a subsequence $\{g_{n_k}\} \subseteq \{g_{n'}\}$ converging uniformly on compacta to an analytic function g, with $g(z_0) = 0$. Since none of the $g_{n'}$'s are zero in Ω, the Hurwitz Theorem (§1.4) implies $g \equiv 0$ in Ω, and consequently $f_{n_k} \to \infty$ uniformly on compact subsets of Ω, as desired.

The proof of the theorem is herewith complete.

The family $\{e^{nz} : n = 1, 2, 3, \ldots\}$ on $\mathbb{C}$ is not normal and shows that the number of exceptional values cannot be reduced to one.

The approach taken here, using the modular function associated with the noneuclidean triangles having zero angles, is a variation due to de la Vallée Poussin [1915].

Remark. If Ω is a domain which includes the point $z = \infty$, then the functions $f(1/z)$ for $f \in \mathcal{F}$ omit a and b in the disk $|z| < 1/R$ for some large R and constitute a normal family of analytic functions there. It follows that $\mathcal{F}$ is normal at ∞, and hence in all of Ω.

Note also that a locally bounded family of analytic functions omits the requisite two points, locally (although not necessarily the same two points), so that the Montel Theorem of §2.2 is subsumed under the FNT by virtue of Theorem 2.1.2.

The FNT can be extended in the following manner (Montel [1912]).

Generalized Normality Test *Suppose that $\mathcal{F}$ is a family of analytic functions in a domain Ω which omit a value $a \in \mathbb{C}$ and such that no function of $\mathcal{F}$ assumes the value $b \in \mathbb{C}$ at more than p points. Then $\mathcal{F}$ is normal in Ω.*

Proof. As in the FNT, we may assume that a and b are 0 and 1, respectively. For $f \in \mathcal{F}$, define $g(z) = \sqrt[p+1]{f(z)}$, taking an arbitrary branch in each case. Then $g(z)$ is analytic in Ω and does not vanish there. Let $w_1, w_2, \ldots, w_{p+1}$ be the $(p+1)$ roots of unity. Suppose that for $z_i \in \Omega$, $g(z_i) = w_i$, for $i = 1, 2, \ldots, p+1$. Then

$$f(z_i) = [g(z_i)]^{p+1} = 1, \qquad i = 1, 2, \ldots, p+1,$$

contradicting the fact that $f(z)$ takes the value 1 at no more than p points. It follows that each function $g(z)$ does not take on a value w_i for some $1 \le i \le p+1$. Then the functions $h(z) = \frac{g(z)}{w_i}$, omit the values 0 and 1, and hence form a normal family. Therefore, the functions $f(z) = [w_i h(z)]^{p+1} = [h(z)]^{p+1}$ comprise a normal family, and the result is proved.

The FNT is readily extended to the setting of meromorphic functions (cf. §3.2), and a further generalization is derived in §4.1, as well as a version for linear fractional transformations in §5.5.

2.8   Picard, Schottky, and Julia Theorems

The following celebrated theorem of E. Picard [1879a] was a revelation in its time and provided an immense stimulus to research in function theory. The original proof used the elliptic modular function, which is also connected with the FNT that is employed in the present proof. In 1896 Borel gave an "elementary" demonstration of Picard's theorem; in other words, a proof which did not utilize the modular function. Bloch [1924] did likewise. Montel [1912] had the incisive idea of replacing a particular property of a function by a family of functions possessing the common property in a sequence of domains. This method was subsequently exploited most efficiently by Julia as well as Montel in a number of different settings which are discussed in the sequel.

Picard's First Theorem *A nonconstant entire function takes every complex value with at most one possible exception.*

Proof. Suppose that $f(z)$ is entire and omits the distinct values $a, b \in \mathbb{C}$. Consider the family of disks $D_n : |z| < 2^n$, $n = 0, 1, 2, \ldots$, and define the functions

$$f_n(z) = f(2^n z),$$

which are also entire. Then $f_n(D_1) = f(D_{n+1})$, $n = 0, 1, 2, \ldots$, and it follows that the sequence $\{f_n\}$ omits the values a and b in D_1. By the FNT (§2.7), $\{f_n\}$ is normal in D_1, and since $f_n(0) = f(0)$, $n = 0, 1, 2, \ldots$, Corollary 2.2.4 implies that $\{f_n\}$ is bounded on the compact subset $\overline{D_0} \subseteq D_1$. We conclude that $f(z)$ is bounded in $\mathbb{C}$, and hence identically constant by Liouville's theorem. This contradiction proves the theorem.

A nonconstant meromorphic function in $\mathbb{C}$ omits at most two values, for if a, b, c were three such values, then the entire function given by $g(z) = (c-b)(f(z)-a)/(c-a)(f(z)-b)$ would omit 0 and 1. The functions e^z, $1/(1 - e^z)$ omit 0 and 0, 1, respectively.

The well-known theorem of Schottky [1904] also derives from the FNT.

Schottky's Theorem *Let f be an analytic function in the disk $|z| < R$ which does not assume the values 0 or 1, and $f(0) = a_0$. Then for each value θ with $0 < \theta < 1$, there exists a positive constant $M(a_0, \theta)$ such that*

$$|f(z)| \leq M(a_0, \theta), \qquad |z| \leq \theta R.$$

Proof. For each such function f satisfying the hypotheses, set

$$\tilde{f}(z) = f(Rz),$$

so that $\tilde{f}$ takes the same values in $|z| < 1$ as does f in $|z| < R$, and $\tilde{f}$ omits the values 0 and 1. Then $\{\tilde{f}\}$ is a normal family satisfying $\tilde{f}(0) = a_0$, and by Corollary 2.2.4, $|\tilde{f}(z)| \leq M$ on any compact subset $|z| \leq \theta < 1$. The bound depends only on θ and a_0, i.e., $M = M(a_0, \theta)$, and we conclude that

$$|f(z)| \leq M(a_0, \theta), \qquad |z| \leq \theta R.$$

Corollary 2.8.1 *Under the conditions of the theorem we obtain*

$$|f(z)| \geq \frac{1}{M\left(\frac{1}{a_0}, \theta\right)}, \quad |f(z) - 1| \geq \frac{1}{M\left(\frac{1}{1-a_0}, \theta\right)}, \quad \frac{1}{|f(z)|} \geq \frac{1}{M(a_0; \theta)},$$

for $|z| \leq \theta R$.

The third inequality is obvious, and the first two follow by considering $\frac{1}{f}$ and $1 - f$ analogously as above. Thus, the theorem places a particular restriction on the growth of the analytic functions which satisfy the hypotheses, the bound depending only on $\theta R = r$ and $f(0)$ (cf. also Theorem 3.6.10).

To take a simple example, every analytic function $f(z)$ with $|f(z)| > 1$ in $|z| < 1$ satisfies the growth condition

$$|f(z)| \leq |f(0)|^{(1+r)/(1-r)}, \qquad 0 \leq |z| = r < 1.$$

This will be deduced in a more general setting in §3.6, but for the reader familar with the Herglotz representation (cf. §5.4) which can be applied directly to $\log|f(z)|$, the inequality becomes transparent. Writing the above as

$$\log M(r, f) \leq \frac{1+r}{1-r} \log|f(0)|,$$

we have a special case of Hayman's [1947] rendering of the Schottky bound:

$$\log M(r, f) \leq \left(\pi + \log^+ |f(0)|\right) \frac{1+r}{1-r},$$

for all $f(z)$ which are analytic in U and omit the values 0 and 1, thus improving upon an earlier estimate of Ahlfors [1938]. Another Schottky type bound has been given by Yang and Chang [1965], generalizing an earlier result of Hiong and Ho [1961].

Theorem 2.8.2 *Let $f(z)$ be an analytic function in U such that all the zeros of $f(z)$ have multiplicity $\geq m$, and all the zeros of $f^{(k)}(z)$ $(k \geq 0; f^{(0)}(z) = f(z))$ have multiplicity $\geq n$, where*

$$\frac{k+1}{m} + \frac{1}{n} < 1.$$

Then

$$\log M(r, f) < \frac{C(k)}{1-r} \left\{ \log^+ |f(0)| + \log \frac{2}{1-r} \right\},$$

where $C(k)$ is a constant depending only on k.

That the set of all such analytic functions as in the preceding theorem is a normal family is a consequence of Theorem 4.4.17 and the subsequent Remark.

Refer to Burckel [1979], p. 431, pp. 456–459 and Hayman [1989], pp.707–708 for other variations on the Schottky theme.

In view of its proof, Schottky's theorem may be rephrased as follows: *Suppose that f is analytic in the $|z| < R$, $|f(0)| \leq \mu$, and does not assume the values 0 and 1. Then for each value θ with $0 < \theta < 1$, there is a positive constant $M = M(\mu, \theta)$ such that*

$$|f(z)| \leq M, \qquad |z| \leq \theta R.$$

A rather immediate proof of the FNT may now be deduced from this version of the Schottky Theorem (which, itself, has an independent derivation in terms of the modular function; cf. Carathéodory [1960, Vol. II], pp. 199–201). Indeed, let $\mathcal{F}$ be a family of analytic functions in a domain Ω, with each $f \in \mathcal{F}$ omitting the values 0 and 1, and choose a point $z_0 \in \Omega$ together with a disk $|z - z_0| < R$ contained in Ω. Partition $\mathcal{F}$ into two distinct subfamilies

$$\mathcal{G} = \{f \in \mathcal{F} : |f(z_0)| \leq 1\} \quad \text{and} \quad \mathcal{H} = \{f \in \mathcal{F} : |f(z_0)| > 1\}.$$

First apply the (rephrased) Schottky Theorem to $f \in \mathcal{G}$ with respect to the disk $|z - z_0| < R$ to deduce that $\mathcal{G}$ is uniformly bounded in a neighbourhood of z_0. A similar argument can then be applied to the function $\frac{1}{f(z)}$ for each $f \in \mathcal{H}$, so that $\frac{1}{f(z)}$ is uniformly bounded in a neighbourhood of z_0, $f \in \mathcal{H}$. It follows that both $\mathcal{G}$ and $\mathcal{H}$ are normal at z_0 (cf. §2.3, No. 9 regarding the normality of $\mathcal{H}$), and likewise for their union $\mathcal{F}$. As z_0 is an arbitrary point of Ω, the FNT is established.

Schottky's theorem can further be used to prove our next result due to Landau [1904], which in turn implies Schottky's theorem.

Landau's Theorem *Let a_0 and $a_1 \neq 0$ be complex numbers. Then there is a constant $M = M(a_0, a_1)$ such that if f is analytic in the disk $|z| < R$, $f(0) = a_0$, $f'(0) = a_1$, and f does not assume the values 0 and 1, then*

$$R \leq M(a_0, a_1).$$

Proof. We apply the Schottky Theorem with $\theta = \frac{1}{2}$, so that $|f(z)| \leq M(a_0)$ in $|z| \leq R/2$, and the bound for f depends only on a_0. By the Cauchy formula,

$$a_1 = \frac{1}{2\pi i} \int_{|\zeta|=R/2} \frac{f(\zeta)d\zeta}{\zeta^2},$$

which implies

$$|a_1| \leq \frac{2M(a_0)}{R} \quad \text{or} \quad R \leq \frac{2M(a_0)}{|a_1|},$$

and this is the desired bound for R.

Landau's theorem may be rephrased in the following way: *For any complex numbers $a_0, a_1 \neq 0$, there exists a constant $M(a_0, a_1)$ such that if $f(z) = a_0 + a_1 z + \dots$ is analytic in $|z| < R$, where $R > M(a_0, a_1)$, then f must assume there at least one of the values 0 or 1.* Refer to Carathéodory [1960, Vol. II], p. 197, for a specific representation of $M(a_0, a_1)$; also §3.6.

For analytic functions $f(z) = a_0 + a_1 z + \dots + a_n z^n + \dots$ of Landau's theorem, if we take $a_n \neq 0$ instead of $a_1 \neq 0$, then one can show in a similar manner that

$$R \leq 2 \sqrt[n]{\frac{M(a_0)}{|a_n|}}.$$

Stronger versions of the Schottky and Landau theorems are dealt with in the next chapter in the context of invariant normal families; harmonic analogues are discussed in §5.4, and a further generalization of Schottky's theorem, established with the aid of quasi-normal families, can be found in the Appendix.

We now prove Picard's Great (Second) Theorem of 1879. Schottky [1904] gave an "elementary" proof based on the ideas of Borel's proof of Picard's First Theorem. The approach here using normal families is, of course, Montel's [1912].

Picard's Great Theorem *If $f(z)$ is analytic in a punctured disk $D = D'(z_0; R)$ about z_0 and z_0 is an essential singularity of $f(z)$, then $f(z)$ attains every finite complex value, infinitely often in D, with at most one possible exception.*

Proof. We may assume $D = \{0 < |z| < R\}$ and let us suppose that there are two values a and b which are omitted by $f(z)$ in D. Then the family of functions $\mathcal{F}$ defined by

$$f_n(z) = f\left(\frac{z}{2^n}\right), \qquad n = 1, 2, 3, \ldots, \tag{2.4}$$

are analytic in the annulus $\mathcal{A} : \frac{R}{2} < |z| < R$, and do not assume the values a or b in $\mathcal{A}$. Since $\mathcal{F}$ is normal in $\mathcal{A}$, there is a subsequence $\{f_{n_k}\}$ converging uniformly to $F(z)$ on the compact set $\{|z| = \rho : \frac{R}{2} < \rho < R\}$, where $F(z)$ is either an analytic function or $\equiv \infty$ in $\mathcal{A}$.

If $F(z)$ is analytic, then its boundedness on $|z| = \rho$ implies $\{f_{n_k}\}$ is uniformly bounded on $|z| = \rho$, say

$$|f_{n_k}(z)| \le M, \qquad |z| = \rho, \;\; k = 1, 2, 3, \ldots .$$

But then

$$|f(z)| \le M, \qquad |z| = \rho/2^{n_k}, \;\; k = 1, 2, 3, \ldots,$$

that is, $f(z)$ is bounded on a sequence of concentric circles converging to the origin. By the Maximum Modulus Theorem, $|f(z)| \le M$ in the region between any two such circles. As a consequence,

$$|f(z)| \le M, \qquad 0 < |z| \le \rho/2^{n_1},$$

which contradicts the fact that $f(z)$ must be unbounded in any neighbourhood of an essential singularity.

If $F(z) \equiv \infty$, then $\left\{\frac{1}{f_{n_k} - a}\right\}$ converges uniformly on compact subsets of $\mathcal{A}$ to zero. As above, we conclude that $h(z) = \frac{1}{f(z)-a}$ is bounded in a deleted neighbourhood of the origin, and thus $f(z)$ has a removable singularity or pole at the origin; again a contradiction.

Finally, if there are two values, say α, β, which are attained only finitely often by $f(z)$, then in some sufficiently small deleted neighbourhood of the origin, $f(z)$ would omit α, β and the result follows.

Observe that $e^{1/z}$ has an essential singularity at the origin, but that $e^{1/z} \neq 0$ for all $z \in \mathbb{C}$, so the above result is best possible.

A meromorphic function omits at most two values in $D = D'(z_0; R)$, since the case of three omitted values can be reduced to consideration of the omitted values $0, 1, \infty$, and the preceding Picard Theorem then applies.

The proof of the theorem suggests the following holds true.

Lemma 2.8.3 *If z_0 is an essential singularity of a function $f(z)$ which is analytic in a punctured neighbourhood $D'(z_0; R)$ of z_0, then the sequence $\{f_n\}$ defined by $f_n(z) = f(\frac{z}{2^n})$ is not normal in, say, $\mathcal{A} : \frac{R}{8} < |z - z_0| < R$.*

Proof. Again, let us take $z_0 = 0$, and suppose that $\{f_n\}$ is normal in $\mathcal{A} : \frac{R}{2^3} < |z| < R$. Then there is a subsequence $\{f_{n_k}\}$ of $\{f_n\}$ that converges normally in $\mathcal{A}$, in particular, uniformly in the closed annulus

$$\overline{\mathcal{A}}_0 : \frac{R}{2^2} \leq |z| \leq \frac{R}{2}.$$

Moreover, the limit function $F(z)$ of $\{f_{n_k}\}$ must be identically infinite, since otherwise, $f(z)$ would be bounded in a deleted neighbourhood of the origin, just as in the proof of the preceding Picard Theorem.

It must also be the case that the sequence $\{f_n\}$ itself converges uniformly on $\overline{\mathcal{A}}_0$ to ∞, for if not, then there exists a sequence of points $z_j \in \overline{\mathcal{A}}_0$, a number $M > 0$, and a subsequence $\{f_{m_j}\}$ of $\{f_n\}$ such that

$$|f_{m_j}(z_j)| \leq M, \qquad j = 1, 2, 3, \ldots .$$

However, this contradicts the fact that $\{f_{m_j}\}$ must contain a subsequence converging uniformly on $\overline{\mathcal{A}}_0$ to ∞; hence the assertion. It follows that for any $M > 0$, there exists an $n_0 = n_0(M)$ such that

$$|f_n(z)| > M, \qquad n \geq n_0, \ z \in \overline{\mathcal{A}}_0.$$

But the values that f_n takes in $\overline{\mathcal{A}}_0$, coincide with the values that f takes in $\overline{\mathcal{A}}_n : \frac{R}{2^{n+2}} \leq |z| \leq \frac{R}{2^{n+1}}$, implying

$$|f(z)| > M, \qquad 0 < |z| \leq \frac{R}{2^{n_0+1}}.$$

In other words, $\lim_{z \to 0} f(z) = \infty$, contradicting the fact that $z = 0$ is an essential singularity.

It was noted by Julia [1924], p. 104, that this result permits the following extension of Picard's theorem.

Julia's Theorem *Let z_0 be an essential singularity of $f(z)$. Then there is at least one ray, $\arg(z - z_0) = \theta$, emanating from z_0, such that in every sector $\theta - \varepsilon < \arg(z - z_0) < \theta + \varepsilon$, $f(z)$ assumes, infinitely often, every complex value with at most one exception.*

Proof. Defining the sequence $\{f_n\}$ as in (2.4), for a punctured neighbourhood D of z_0 (which we again take to be the origin), Lemma 2.8.3 implies that $\{f_n\}$ is not normal at some point $\zeta_0 \in D$, and hence not normal in an arbitrarily small disk $D_0 : |z - \zeta_0| < r$. Then, with regard to the family of homothetic disks

$$D_n : \left| z - \frac{\zeta_0}{2^n} \right| < \frac{r}{2^n}, \qquad n = 1, 2, 3, \ldots,$$

we see that $f_n(D_0) = f(D_n)$.

Given any two values $a, b \in \mathbb{C}$, suppose that it is not the case at least one of them is attained in infinitely many disks D_n. Then there is some $n_0 \in \mathbb{N}$ such that for $n \geq n_0$, a and b are not attained by f in D_n, that is, a and b are not attained by f_n in D_0, $n \geq n_0$. By the FNT (§2.7), $\{f_n\}$ is normal in D_0, a contradiction, and the theorem follows.

The ray issuing from the essential singularity z_0 and passing through ζ_0 is called the *direction of Julia*. Refer to §3.7 and §5.4 for other aspects of Julia's theorem.

Consideration of the family

$$f_n(z) = f(2^n z), \qquad n = 1, 2, 3, \ldots,$$

for $f(z)$ an entire function which is not a polynomial yields analogously as above that $f(z)$ assumes every value infinitely often with at most one possible exception in the sector $\theta - \varepsilon < \arg z < \theta + \varepsilon$.

2.9 Sectorial Theorems

Early on, Montel applied the ideas of normal families to various sectorial theorems, inspired by the work of Lindelöf [1909]. The following is sometimes also referred to as *Montel's Theorem* [1912] in the literature; cf. also Lindelöf [1915] for a similar result.

Theorem 2.9.1 *Let $f(z)$ be analytic in the sector $S = \{z : 0 < |z| < R,$ $\alpha < \arg z < \beta\}$, and suppose that $f(z)$ omits two values. If $f(z) \to \ell \in \widehat{\mathbb{C}}$ as $z \to 0$ along some ray $\arg z = \gamma$, where $\alpha < \gamma < \beta$, then $f(z) \to \ell$ as $z \to 0$ uniformly in $\alpha + \delta \leq \arg z \leq \beta - \delta$, for any $0 < \delta < \beta - \alpha$.*

Proof. Assume first that $\ell \in \mathbb{C}$ and define

$$S_n = \left\{ z : \frac{R}{2^{n+2}} < |z| < \frac{3R}{2^{n+2}}, \ \alpha < \arg z < \beta \right\}, \qquad n = 0, 1, 2, \ldots,$$

as well as functions

$$f_n(z) = f\left(\frac{z}{2^n}\right), \qquad z \in S_0.$$

Then $f_n(S_0) = f(S_n)$, $n = 1, 2, 3, \ldots$, and since $f(z)$ omits two values, $\{f_n\}$ is normal in S_0. We can now apply the Vitali-Porter Theorem (§2.4) since $\lim_{n\to\infty} f_n(z) = \ell$ on the segment $\arg z = \gamma$, $\frac{R}{4} < |z| < \frac{3R}{4}$. Hence $f_n(z) \to \ell$ uniformly on compact subsets of S_0. Without loss of generality, we may assume $f_n \to \ell$ uniformly on the compact set

$$K = \left\{ z : \frac{R}{4} \le |z| \le \frac{3R}{4}, \ \alpha + \delta \le \arg z \le \beta - \delta \right\}.$$

Then given $\varepsilon > 0$, there exists $n_0 = n_0(\varepsilon, K)$ such that

$$|f_n(z) - \ell| < \varepsilon, \qquad n \ge n_0, \ z \in K. \tag{2.5}$$

We conclude that if $|\zeta| \le \frac{3R}{2^{n_0+2}}$, $\alpha + \delta \le \arg \zeta \le \beta - \delta$, then (2.5) implies

$$|f(\zeta) - \ell| < \varepsilon,$$

as desired.

If $\ell = \infty$, we can apply a similar argument to the function $1/f(z)$ to obtain the conclusion.

Similar reasoning can be applied to bounded analytic functions in the open unit disk to demonstrate that such functions possess an *angular limit* at a boundary point, which is induced by the existence of a *radial limit* at the point (cf. Duren [1970], p. 6). This issue is dealt with more generally in §5.3 in the context of *normal functions*.

There is of course an analogous result to that of the theorem for an open sector $S' = \{z : |z| > 0, \ \alpha < \arg z < \beta\}$, in which $f(z) \to \ell$ as $z \to \infty$ along a ray in the sector (cf. Burckel [1979], p. 441). A generalization of the theorem has also been obtained by Cartwright [1935a] based on an adaptation of Montel's proof.

The following theorem of Lindelöf [1909] is a direct consequence of Theorem 2.9.1.

Corollary 2.9.2 *Let $f(z)$ be analytic in the sector S and suppose that there are two rays $\arg z = \gamma_1$, $\arg z = \gamma_2$ with $\alpha < \gamma_1 < \gamma_2 < \beta$ such that*

$$\lim_{r\to 0} f(re^{i\gamma_1}) = c_1, \quad \lim_{r\to 0} f(re^{i\gamma_2}) = c_2.$$

Then either $c_1 = c_2$ or $f(z)$ omits at most one value in the sector $\gamma_1 \le \arg z \le \gamma_2$.

Since the sector S of Theorem 2.9.1 is conformally equivalent to an infinite half-strip, Montel's theorem can be rephrased as

Theorem 2.9.3 *Assume $f(z)$, $z = x + iy$, is analytic in the half-strip $S : a < x < b, \ y > 0$, and $f(z)$ omits two values for $z \in S$. If $f(z) \to \ell$ as $y \to \infty$ for a fixed $x = x_0$, then $f(z) \to \ell$ on every line $x = \xi$ in S, and indeed $f(z) \to \ell$ as $y \to \infty$ uniformly for $a + \delta \le x \le b - \delta$.*

The analogue of Montel's theorem for $|f(z)|$ is in general false, since the function $f(z) = e^{\sinh z}$ is bounded in any half-strip S as above, but $|f(x+iy)|$ approaches a limit as $y \to \infty$ if and only if $x = 0$. However, Hardy, Ingham, and Pólya [1928] showed that under certain conditions an analogue of Montel's theorem holds if $|f(x+iy)|$ approaches a limit as $y \to \infty$ for *two distinct values* of x. (This version has a harmonic counterpart in the Hardy-Montel Theorem proved in §5.4). Cartwright [1962] showed that convergence along only one line was sufficient under the following circumstances: *In S, if $f(z) \ne 0$, $|f(z)| < 1$, and $|f(x_0 + iy)| \to 1$ as $y \to \infty$, then $|f(x+iy)| \to 1$ as $y \to \infty$ uniformly for $a + \delta \le x \le b - \delta$.* Compare also Bohr [1927], Hayman [1962], Bowen [1964–65], as well as Burckel [1979], Chapter 12, for other versions. Generalizations to abstract Riemann surfaces have been obtained by Ohtsuka [1956].

2.10   Covering Theorems

For any function f belonging to the *schlicht* class $\mathcal{S}$, the Koebe One-Quarter Theorem states that the image of the unit disk, $f(U)$, entirely covers the disk $|w| < \frac{1}{4}$ (cf. Burckel [1979], p. 232). In this regard, the requirement of univalence is essential, for the function

$$f(z) = \frac{1 - (1-z)^n}{n}, \qquad n > 0,$$

is analytic in U, with $f(0) = 0$, $f'(0) = 1$. However, $f(z) \ne \frac{1}{n}$ for any $z \in U$, and $\frac{1}{n}$ can be chosen arbitrarily close to the origin. Hence, for the class

$$\mathcal{T} = \{f : f \text{ analytic in } U, \ f(0) = 0, \ f'(0) = 1\},$$

we cannot expect to find a similar covering theorem to that of Koebe. However, somewhat weaker versions can be demonstrated by normal family arguments. Our first such result in this direction is due to Valiron [1927].

Theorem 2.10.1 *Given $\varepsilon \in (0, 2\pi)$, there exists $\rho_\varepsilon > 0$ such that for each $f \in \mathcal{T}$, $f(U)$ covers a circular sector centred at the origin of radius ρ_ε and central angle $2\pi - \varepsilon$.*

Proof. It suffices to consider only the case $\varepsilon < \pi$. Suppose that there is no such ρ_ε as claimed in the theorem. Then we may find sequences $\{a_n\}$ and $\{b_n\}$ in $\mathbb{C}$ such that $a_n \to 0$, $b_n \to 0$, as $n \to \infty$, with the difference of their arguments $\phi_n = \arg b_n - \arg a_n$ satisfying

$$\varepsilon < \phi_n < \pi, \qquad n = 1, 2, 3, \ldots,$$

together with a sequence of functions $\{f_n\} \subseteq \mathcal{T}$ that satisfy $f_n(z) \ne a_n$, $f_n(z) \ne b_n$, for all $z \in U$, $n = 1, 2, 3, \ldots$. Firstly, suppose that $\phi_n < \frac{\pi}{2}$. If we drop a line segment from the point a_n that is perpendicular to the line from the origin through the point b_n, then clearly

$$|b_n - a_n| \geq |a_n| \sin \phi_n,$$

the latter term being the length of the perpendicular line segment. It follows that $|b_n - a_n| \geq |a_n| \sin \varepsilon$, and so

$$\frac{|a_n|}{|b_n - a_n|} \leq \frac{1}{\sin \varepsilon}.$$

If, however, $\phi_n \geq \frac{\pi}{2}$, by implementing the same construction as above, we deduce that $|b_n - a_n| \geq |a_n|$, and therefore

$$\frac{|a_n|}{|b_n - a_n|} \leq 1 \leq \frac{1}{\sin \varepsilon}.$$

Next, define the functions

$$g_n(z) = \frac{f_n(z) - a_n}{b_n - a_n}, \qquad z \in U, \ n = 1, 2, 3 \ldots .$$

Since the $g_n(z)$ are analytic in U and do not take the values 0 or 1 there, the family $\{g_n\}$ is normal in U. Moreover,

$$|g_n(0)| \leq \frac{1}{\sin \varepsilon}, \qquad n = 1, 2, 3 \ldots ,$$

from the above. As a consequence, there is a subsequence $\{g_{n_k}\}$ that converges uniformly on compact subsets of U to an analytic function $g(z)$. Now, $g'_{n_k}(0) \to g'(0)$, in other words

$$\frac{1}{b_{n_k} - a_{n_k}} \to g'(0),$$

which violates the fact that a_{n_k} and b_{n_k} tend to zero, and the theorem is proved.

The best value for ρ_ε, given ε, was determined by Bermant and Lavrentieff [1935]. Another version of Valiron's theorem can be found in Yang and Chang [1965].

By a means similar to that in the proof of Valiron's theorem, we have the (somewhat remarkable) result of Montel [1933], pp. 118-122.

Theorem 2.10.2 *Let $\mathcal{A}$ denote the annulus $0 < R_1 < |w| < R_2$, and let p be an arbitrary positive integer. Then there exists $\rho = \rho(R_1, R_2) > 0$ such that for $f \in \mathcal{T}$, either $f(U) \supseteq \{|w| < \rho\}$, or $f(z)$ assumes every value of $\mathcal{A}$ at no less than p points in U.*

Proof. Suppose not. Then there are points $\{a_n\}$, $\{b_n\}$, such that $a_n \to 0$ as $n \to \infty$ and $b_n \in \mathcal{A}$, and a sequence $\{f_n\} \subseteq \mathcal{T}$ such that for $z \in U$

$$f_n(z) \neq a_n,$$
$$f_n(z) = b_n$$

for less than p values of z. Define

$$g_n(z) = \frac{f_n(z) - a_n}{b_n - a_n}, \qquad z \in U, \ n = 1, 2, 3, \ldots .$$

The family $\{g_n\}$ does not assume the value 0 in U, and assumes the value 1 at no more than $(p-1)$ values of $z \in U$. By the Generalized Normality Test (§2.7), $\{g_n\}$ is a normal family with $|g_n(0)| = \frac{|a_n|}{|b_n - a_n|} \leq M < \infty$, $n = 1, 2, 3, \ldots$. Therefore, there is a subsequence $\{g_{n_k}\}$ which converges uniformly on compact subsets of U to an analytic function g, with

$$g(0) = \lim_{k \to \infty} \frac{-a_{n_k}}{b_{n_k} - a_{n_k}} = 0.$$

Since the origin cannot be a limit point of zeros of the g_{n_k}, we conclude that $g \equiv 0$ in U. But then $g'_{n_k}(0) = \frac{1}{b_{n_k} - a_{n_k}}$ should tend to zero as $k \to \infty$, which it clearly does not, and the theorem is proved.

The following variation of the preceding theorem was demonstrated by Yang [1964] and Yang and Chang [1965] using a similar argument; cf. also the Remark following Theorem 4.4.17.

Theorem 2.10.3 *Let $\mathcal{A}$ denote the annulus $0 < R_1 < |w| < R_2$, and let q be an arbitrary positive integer. Then there exists $\rho = \rho(R_1, R_2) > 0$, such that for any function $f \in \mathcal{T}$ either one of the following obtains:*

(i) *For each point w of the disk $|w| < \rho$, the equation $f(z) = w$ has at least one root in U of multiplicity less than $2q + 3$.*

(ii) *For each $w \in \mathcal{A}$, the equation $f^{(k)}(z) = w$ $\big(0 \leq k \leq q;\ f^{(0)}(z) = f(z)\big)$ has at least one simple root in U.*

By virtue of Theorems 2.10.2 or 2.10.3 we maintain:

Corollary 2.10.4 *Let $\mathcal{T}_\zeta$ be the family of $f \in \mathcal{T}$ which omit some given nonzero value $\zeta \in \mathbb{C}$. Then there is some $\rho > 0$ such that for all $f \in \mathcal{T}_\zeta$, $f(U) \supseteq \{|w| < \rho\}$.*

In fact, taking the annulus $\mathcal{A}$ above to contain the point ζ proves the theorem.

Another immediate consequence of the two preceding theorems is due to Landau [1922].

Corollary 2.10.5 *There exists a $\rho > 0$ such that for each $f \in \mathcal{T}$, $f(U)$ entirely covers some circumference of radius at least ρ.*

In Hayman [1989], pp. 452–453, there is a sharpening of the Landau result: *If $f \in \mathcal{T}$ and $\varepsilon > 0$, then f attains in U all the values on some circle of radius $r \geq \frac{1}{4} - \varepsilon$.* As the Koebe function

$$k(z) = \frac{z}{(1-z)^2}$$

belongs to $\mathcal{T}$ and omits the real values $w \leq -\frac{1}{4}$, the ε cannot be taken equal to zero. The preceding may be construed as the analogue of the One-Quarter Theorem for the class $\mathcal{T}$, and is actually a special case of a much more general such theorem (Hayman [1989], Theorem 7.24).

Extensions of the covering theorems of Montel, Landau, as well as Corollary 2.10.4, were obtained by Bermant [1944] in his deep investigation of the modular function.

Finally in this vein, we have a theorem of Fekete [1925].

Theorem 2.10.6 *Let $\mathcal{T}_p$ be the family of all $f \in \mathcal{T}$ which have no more than p simple zeros in U. Then there is a $\rho > 0$ such that for all $f \in \mathcal{T}_p$, $f(U) \supseteq \{|w| < \rho\}$.*

Proof. Suppose that the conclusion does not hold. Then there exists a sequence of nonzero points $\{w_n\}$ with $w_n \to 0$ as $n \to \infty$, and a sequence $\{f_n\} \subseteq \mathcal{T}_p$ such that $f_n(z) \neq w_n$, for $z \in U$, $n = 1, 2, 3, \ldots$. The functions $g_n(z) = \frac{f_n(z)}{w_n}$ have no more than p simple zeros in U, and the sequence $\{g_n\}$ omits the value 1. Then $\{g_n\}$ is normal by the Generalized Normality Test (§2.7) and contains a subsequence $\{g_{n_k}\}$ converging normally to an analytic function g, since $g_n(0) = 0$, $n = 1, 2, 3, \ldots$. But then $g'_{n_k}(0) = \frac{1}{w_{n_k}} \to g'(0)$, a contradiction, since $\frac{1}{w_{n_k}} \to \infty$ as $k \to \infty$.

In the case $p = 1$, we retrieve a theorem of Carathéodory [1907] who obtained the value $\rho = \frac{1}{16}$ for the radius of the covered disk.

Other covering theorems due to Bloch and Landau are discussed in §4.3.

2.11   Normal Convergence of Univalent Functions

Given a sequence $\{f_n\}$ of analytic univalent functions defined on the open unit disk U satisfying $f_n(0) = 0$, $f'_n(0) > 0$, denote $\Omega_n = f_n(U)$, for $n = 1, 2, 3, \ldots$. In 1912, Carathéodory gave a complete characterization of the normal convergence of $\{f_n\}$ solely in terms of the convergence of their image domains Ω_n. Observe that if such f_n converge normally to f in U, then f is either univalent or the constant zero by virtue of Corollary 1.4.1.

A key ingredient in Carathéodory's work is the following concept. Let $\{\Omega_n\}$ be a sequence of domains in $\mathbb{C}$ with $0 \in \Omega_n$. Then the *kernel of*

$\{\Omega_n\}$ *with respect to* 0, $\ker\{\Omega_n\}$, is defined as follows. If some disk $D(0;\rho)$, $\rho > 0$ is contained in all the domains Ω_n, then $\ker\{\Omega_n\}$ is the largest domain containing 0 every compact subset of which is contained in all but finitely many of the Ω_n; if no such disk exists, then $\ker\{\Omega_n\} = \{0\}$. By "largest" is meant that any other domain with this property is contained in the kernel. It is worth noting that $\ker\{\Omega_n\}$ may be $\mathbb{C}$. In addition, we say that $\{\Omega_n\}$ *converges to* $\Omega_0 = \ker\{\Omega_n\}$, written $\Omega_n \to \Omega_0$, if every subsequence of these domains has Ω_0 as is kernel. For example, let Ω_{2k} be the open unit disk U, and Ω_{2k+1} be the open unit square S (centred at the origin). Then $\ker\{\Omega_n\} = U$, whereas $\ker\{\Omega_{2k+1}\} = S$, so that Ω_n does not converge to the kernel.

The case when $\ker\{\Omega_n\} = \{0\}$ is readily dealt with. Indeed, suppose that $f_n \to f$ normally in U. As there is no $r > 0$ such that $K(0;r) \subseteq \Omega_n$ for all n sufficiently large, then $\operatorname{dist}(0, \delta\Omega_n) \to 0$ as $n \to \infty$. An application of the growth condition of Example 5 (§2.3) to the analytic univalent function $\frac{f_n(z)}{f_n'(0)} \in \mathcal{S}$ gives

$$|f_n(z)| \le \frac{|z|}{(1 - |z|)^2}|f_n'(0)|.$$

Coupled with the Koebe Distortion Theorem (cf. Pommerenke [1975], Corollary 1.4): $\frac{1}{4}|f_n'(0)| \le \operatorname{dist}(0, \delta\Omega_n)$, yields the result $f \equiv 0$.

On the other hand, if $f_n \to 0$ normally in U, then $\ker\{\Omega_n\}$ contains no disk $D(0;r)$, $r > 0$. For if it did, then the argument of part (ii) below would give $f_n'(0) \ge \frac{1}{r} > 0$, contradicting the normal convergence of f_n to 0. Therefore, $\ker\{\Omega_n\} = \{0\}$.

We now demonstrate the

Carathéodory Convergence Theorem *Let* $\{f_n\}$ *be a sequence of analytic univalent functions on U with $f_n(0) = 0$, $f_n'(0) > 0$, and $\Omega_n = f_n(U)$, $n = 1, 2, 3, \ldots$. Then f_n converges normally in U to a univalent function f if and only if $\Omega_0 = \ker\{\Omega_n\} \ne \{0\}$, $\mathbb{C}$, and $\Omega_n \to \Omega_0$. Moreover, $\Omega_0 = f(U)$.*

Proof. (i) Suppose that f_n converges normally in U to a univalent function f. By the preceding discussion, $\Omega_0 = \ker\{\Omega_n\} \ne \{0\}$. We aim to show that $f(U) = \Omega_0$. This being the case, it immediately follows that $\Omega_0 \ne \mathbb{C}$ since, if it were equal to $\mathbb{C}$, the inverse function $f^{-1} : \mathbb{C} \to U$ would violate Liouville's theorem. Moreover, as every subsequence of $\{f_n\}$ converges normally to f, it transpires that every subsequence of $\{\Omega_n\}$ has kernel $= \Omega_0$, i.e., $\Omega_n \to \Omega_0$.

We first prove: $f(U) \subseteq \Omega_0$. In fact, given any compact subset $K \subseteq f(U)$, there exists $0 < r < 1$ with $K \subseteq f(D(0;r))$. Setting $\Gamma = f(\{|z| = r\})$, we have $\operatorname{dist}(K, \Gamma) = d > 0$ as K and Γ are disjoint compact sets. Since $f_n \to f$ normally in U, there exists $n_0 = n_0(K)$ such that

$$|f_n(re^{i\theta}) - f(re^{i\theta})| < \frac{d}{2}$$

for all $n > n_0$. Now both Γ and $\Gamma_n = f_n(\{|z| = r\})$ have winding numbers $= +1$ about K, so that by the Argument Principle, each value in K is attained exactly once by f_n, i.e., $K \subseteq \Omega_n = f_n(U)$, $n > n_0$. Since $0 \in f(U)$, the result follows from the definition of Ω_0.

To show the opposite inclusion $\Omega_0 \subseteq f(U)$, we take an arbitrary point $0 \neq w_0 \in \Omega_0$ and a closed disk $K(w_0; r) \subseteq \Omega_0$ together with a closed disk $K(0; \rho) \subseteq \Omega_0$ and set $K = K(w_0; r) \cup K(0; \rho)$. Then $K \subseteq \Omega_n$ for all $n > n_0$, say, so that $\phi_n = f_n^{-1}$ is defined on K, $n > n_0$, and satisfies $|\phi_n(w)| < 1$, $\phi_n(0) = 0$. Therefore, $\{\phi_n\}$ is a normal family (in K^0) which has a subsequence $\{\phi_{n_k}\}$ that converges normally in K^0 to ϕ, where ϕ is analytic (indeed univalent), with $\phi(0) = 0$, $|\phi(w)| \leq 1$. Hence for $w \in K^0$, $|\phi(w)| < 1$ by the Maximum Modulus Theorem and so $\phi(w_0) = z_0 \in U$. Since $f_{n_k} \to f$ uniformly in a neighbourhood of z_0, we see that

$$\phi_{n_k}(w_0) \to \phi(w_0) \quad \text{implies} \quad f_{n_k}(\phi_{n_k}(w_0)) \to f(\phi(w_0)),$$

that is, $w_0 = f(z_0) \in f(U)$ as desired.

(ii) To establish the converse, suppose that $\Omega_0 \neq \{0\}$, $\mathbb{C}$, and that $\Omega_n \to \Omega_0$. If $D(0; r) \subseteq \Omega_n$ for all n, then the functions $\phi_n = f_n^{-1}$ are defined on $D(0; r)$ for all n. Consider the functions $\psi_n : U \to U$ given by

$$\psi_n(w) = \phi_n(rw).$$

Schwarz's lemma (§1.4) may then be applied, since $\psi_n(0) = 0$, to give

$$r_n \phi_n'(0) = \psi_n'(0) \leq 1,$$

i.e., $\phi_n'(0) \leq \frac{1}{r}$. Therefore, $f_n'(0) \geq r > 0$, and $\{f_n'(0)\}$ is bounded below.

Furthermore, we claim that $\{f_n'(0)\}$ is also bounded above. For if not, then some subsequence, again denoted by $\{f_n'(0)\}$, converges to $+\infty$. Moreover ,

$$\text{dist}(0, \delta\Omega_n) \geq \frac{f_n'(0)}{4},$$

as above, which means that $D\left(0; \frac{f_n'(0)}{4}\right) \subseteq \Omega_n$, $n = 1, 2, 3, \ldots$. However, this leads to the contradiction $\Omega_0 = \mathbb{C}$, and so $\{f_n'(0)\}$ is bounded above as asserted. As a consequence, $\{f_n\}$ is locally bounded, hence normal (cf. the discussion after Example 10 of §2.3), so that some subsequence $\{f_{n_k}\}$ converges normally in U to a univalent function f (since the kernel $\neq \{0\}$), with $f(0) = 0$, $0 < r \leq f'(0) < \infty$.

To complete the proof, we must show that $\{f_n\}$ itself converges normally in U to f. If this were not the case, then as in the proof of the Vitali-Porter Theorem (§2.4), there would be another subsequence $\{f_{m_k}\}$ that converges normally in U to a limit function g distinct from f. From the proof of part (i) we obtain $\ker\{\Omega_{n_k}\} = f(U)$, $\ker\{\Omega_{m_k}\} = g(U)$. Since $\Omega_n \to \Omega_0$, $\Omega_0 = f(U) = g(U)$ obtains. Furthermore, $f(0) = g(0) = 0$ and $f'(0) > 0$, $g'(0) > 0$, implying $f(z) \equiv g(z)$ by the uniqueness of the Riemann mapping function (§2.6). This contradiction establishes that $f_n \to f$ normally in U as desired.

For generalizations of Carathéodory's theorem, see Goluzin [1969], p. 59; see also Pfluger [1969], pp. 184–186, for some further examples. By way of contrast with Osgood's theorem (§2.4) we have

Corollary 2.11.1 *Let $\{f_n\}$ be a sequence of analytic univalent functions in U, $f_n(0) = 0$, $f'_n(0) > 0$, such that f_n converges pointwise to an analytic univalent function f. Then $\{f_n\}$ converges uniformly on compact subsets of U to f.*

Proof. Applying the standard growth estimate

$$f'_n(0) \frac{|z|}{(1+|z|)^2} \le |f_n(z)| \le \frac{|z|}{(1-|z|)^2} f'_n(0)$$

to any nonzero $z_0 \in U$ gives $f'_n(0) \le \frac{(1+|z_0|)^2}{|z_0|} |f_n(z_0)|$, and we conclude that $\{f'_n(0)\}$ is bounded above. Similarly it is deduced that $\{f'_n(0)\}$ is bounded away from zero. This once again assures the local boundedness of $\{f_n\}$ and the result is now a consequence of the Vitali-Porter Theorem.

3

Meromorphic Functions

Normal families of meromorphic functions are most naturally studied using the spherical metric (§1.2), an approach initiated by Ostrowski [1926]. Some results for meromorphic functions, such as the FNT, are immediate extensions from the analytic case, whereas others, such as Landau's or Julia's theorem are set in a much broader context than their analytic counterparts. Normality criteria more pertinent to families of meromorphic functions, such as Marty's theorem, have not yet been encountered. The notion of normal invariant families is introduced, and these play a further role in the discussion of normal functions in Chapter 5. One of the highlights of the normal invariant theory is a condition for normality based on the Ahlfors Five Islands Theorem. We conclude with some observations on the normality of families of linear fractional transformations which are again featured in Chapter 5 in the context of discontinuous groups. Additional results involving meromorphic functions are also in the next chapter on the Bloch Principle.

3.1 Normality

We begin with

Definition 3.1.1 *A family $\mathcal{F}$ of functions meromorphic in a domain Ω is* **normal** *in Ω if every sequence $\{f_n\} \subseteq \mathcal{F}$ contains a subsequence which converges spherically uniformly on compact subsets of Ω.*

That the limit function is either meromorphic in Ω or identically equal to ∞ is a consequence of Corollary 3.1.4. That the limit function can actually be identically ∞ is given by

Example 3.1.2 Let $f_n(z) = \frac{n}{z}$, $n = 1, 2, 3, \ldots$, on U. Then each f_n is meromorphic and $\{f_n\}$ converges spherically uniformly to ∞ in U.

In order to relate normal convergence in the chordal metric with uniform convergence in the euclidean metric, the next result is frequently used in the sequel.

Theorem 3.1.3 *Let $\{f_n\}$ be a sequence of meromorphic functions on a domain Ω. Then $\{f_n\}$ converges spherically uniformly on compact subsets of Ω to f if and only if about each point $z_0 \in \Omega$ there is a closed disk $K(z_0; r)$ in which*

$$|f_n - f| \to 0$$

or

$$\left| \frac{1}{f_n} - \frac{1}{f} \right| \to 0,$$

uniformly as $n \to \infty$.

Proof. Since $\chi(w_1, w_2) \leq |w_1 - w_2|$ and $\chi(w_1, w_2) \leq \left| \frac{1}{w_1} - \frac{1}{w_2} \right|$, if either $|f_n - f| \to 0$ or $\left| \frac{1}{f_n} - \frac{1}{f} \right| \to 0$ uniformly in $K(z_0; r)$, then $\chi(f_n, f) \to 0$ uniformly in $K(z_0; r)$, and hence on compact subsets of Ω.

On the other hand, suppose that $\chi(f_n, f) \to 0$ uniformly on compact subsets of Ω. There are two cases.

Case (i): $f(z_0) \neq \infty$.

In view of Proposition 1.6.2, $f(z)$ is spherically continuous in Ω, hence there is some closed neighbourhood $K(z_0; r)$ on which $f(z)$ is bounded. Therefore, by Theorem 1.2.2, $f_n \to f$ uniformly on $K(z_0; r)$. Note that $f(z)$ is analytic in the interior of $K(z_0; r)$.

Case (ii): $f(z_0) = \infty$.

As above, we can find a closed disk $K(z_0; r)$ in which $\frac{1}{f(z)}$ is bounded. Since $\chi\left(\frac{1}{f_n}, \frac{1}{f} \right) \to 0$ uniformly in this disk, again by Theorem 1.2.2, $\frac{1}{f_n} \to \frac{1}{f}$ uniformly in $K(z_0; r)$.

Corollary 3.1.4 *Let $\{f_n\}$ be a sequence of meromorphic functions on Ω which converges spherically uniformly on compact subsets to f. Then f is either a meromorphic function on Ω or identically equal to ∞.*

Proof. Suppose that $f \not\equiv \infty$. Then for some $z \in \Omega$, $f(z) \neq \infty$, and according to Case (i), f is analytic in a neighbourhood of every point where it is finite.

On the other hand, if $f(z_0) = \infty$ ($f \not\equiv \infty$), we claim z_0 is an isolated singularity. For if not, let $\{z_n\}$ be a sequence in Ω such that $z_n \to z_0$, $f(z_n) = \infty$. By Case (ii), $\frac{1}{f}$ is analytic in $D(z_0; r)$. Then $\frac{1}{f(z_n)} = 0$ for all n, implying $\frac{1}{f} \equiv 0$ in $D(z_0; r)$, i.e., $f \equiv \infty$ in $D(z_0; r)$. Let

$$S = \{z \in \Omega : f(z) = \infty\} = f^{-1}(\infty).$$

Then $S \neq \emptyset$ and is an open set by the above. Moreover, S is trivially a closed set. The connectedness of Ω then implies $S \equiv \Omega$, contradicting $f \not\equiv \infty$, so that z_0 must be an isolated singularity and f is meromorphic in Ω.

Corollary 3.1.5 *Let $\{f_n\}$ be a sequence of analytic functions on a domain Ω which converge spherically uniformly on compact subsets of Ω to f. Then f is either analytic on Ω or identically equal to ∞.*

Proof. Suppose that $f(z_0) = \infty$ for some $z_0 \in \Omega$. Since $\frac{1}{f_n} \to \frac{1}{f}$ uniformly in some $D(z_0;r)$, $\frac{1}{f_n}$ is analytic, $\neq 0$ in $D(z_0;r)$ for all sufficiently large n. So it must be that $\frac{1}{f} \equiv 0$ in $D(z_0;r)$ in view of Hurwitz's theorem (§1.4), i.e., $f \equiv \infty$ there. Then $f \equiv \infty$ as in the preceding corollary.

If $f \not\equiv \infty$, the conclusion follows also as in the preceding corollary.

In order to show the equivalence of the two definitions of normality for a family of analytic functions, we maintain

Proposition 3.1.6 *A sequence $\{f_n\}$ of analytic functions on a domain Ω converges uniformly on compact subsets to f (which may be $\equiv \infty$) if and only if $\{f_n\}$ converges spherically uniformly on compact subsets to f.*

Indeed, uniform convergence implies spherical uniform convergence whenever the limit function is analytic, in which case the converse is also obvious from Theorem 1.2.2. When $f \equiv \infty$, the relation

$$\chi(f_n, \infty) = \frac{1}{\sqrt{1 + |f_n|^2}}$$

yields the assertion.

In view of Definitions 2.1.1 and 3.1.1, we have

Corollary 3.1.7 *A family $\mathcal{F}$ of analytic functions is normal in a domain Ω with respect to the usual metric if and only if $\mathcal{F}$ is normal in Ω with respect to the chordal metric.*

With regard to derivatives, we know (by Corollary 2.2.5) that if $\mathcal{F}$ is a normal family of analytic functions whose limit functions are all analytic, then the family $\mathcal{F}'$ of derivatives constitutes a normal family as well. However, this turns out not to be the case for meromorphic functions.

Example 3.1.8 Define

$$f_n(z) = \frac{n^2}{1 - n^2 z^2} = \frac{1}{\frac{1}{n^2} - z^2}, \qquad n = 1, 2, 3, \ldots .$$

Then $f_n(z)$ converges to $-\frac{1}{z^2} = f(z)$ normally in $\mathbb{C}$, resulting in the normality of $\{f_n\}$. On the other hand,

$$f_n'(z) = \frac{2n^4 z}{(1 - n^2 z^2)^2} = \frac{2z}{\left(\frac{1}{n^2} - z^2\right)^2}$$

satisfies $f_n'(0) = 0$, whereas the limit function ($z \neq 0$), $F(z) = \frac{2}{z^3}$, is ∞ at $z = 0$. We conclude that $\{f_n'\}$ is not normal in any neighbourhood of the origin.

3.2   Montel's Theorem

In discussing the normality of a family of meromorphic functions, the concept of local boundedness is not entirely relevant (although see Corollary 3.3.4 where it is partially resurrected). However, spherical equicontinuity can be substituted in the following counterpart to Montel's theorem (Ostrowski [1926]):

Theorem 3.2.1 *A family $\mathcal{F}$ of meromorphic functions in a domain Ω is normal if and only if $\mathcal{F}$ is spherically equicontinuous in Ω.*

Proof. Suppose that $\mathcal{F}$ is normal but not spherically equicontinuous. Then there is a point $z_0 \in \Omega$, some $\varepsilon > 0$, a sequence $\{z_n\} \subseteq \Omega$ with $z_n \to z_0$, and a sequence $\{f_n\} \subseteq \mathcal{F}$ satisfying

$$\chi\big(f_n(z_0), f_n(z_n)\big) > \varepsilon, \qquad n = 1, 2, 3, \dots . \tag{3.1}$$

The normality of $\mathcal{F}$ provides a subsequence $\{f_{n_k}\}$ converging spherically uniformly on compact subsets; in particular on a compact subset containing $\{z_n\}$. But according to Proposition 1.6.2, $\{f_{n_k}\}$ is spherically equicontinuous at z_0, violating (3.1). Therefore, $\mathcal{F}$ is spherically equicontinuous.

The converse is deduced in the same manner, *mutatis mutandis*, as Montel's theorem (§2.2), owing to the compactness of the Riemann sphere.

Normality can also be studied from the equivalent (but no longer fashionable) point of view of *spherical oscillation*, an approach conceived by Ostrowski [1926] and also considered by Carathéodory [1929]. For an explication of the basic ideas, as well as the related notion of *continuous convergence*, see Carathéodory [1960, Vol. I], pp. 173–183.

The Fundamental Normality Test also has the following analogue for meromorphic functions.

Fundamental Normality Test   *Let $\mathcal{F}$ be a family of meromorphic functions on a domain Ω which omit three distinct values a, b, c in $\mathbb{C}$. Then $\mathcal{F}$ is normal in Ω.*

Proof. Set

$$\mathcal{G} = \left\{ g(z) = \frac{c-a}{c-b} \cdot \frac{f(z)-b}{f(z)-a} : f \in \mathcal{F} \right\}.$$

Then $\mathcal{G}$ is a family of analytic functions in Ω which omit the values 0 and 1. Hence by the analytic FNT (§2.7), $\mathcal{G}$ is normal in the usual metric, and consequently in the chordal metric. The normality of $\mathcal{F}$ then follows.

The FNT also holds for $\Omega \subseteq \widehat{\mathbb{C}}$; see comments after Theorem 3.3.2. Montel [1916] gave a generalization of the FNT for meromorphic functions, namely

Theorem 3.2.2 *Let $\mathcal{F}$ be a family of meromorphic functions in Ω for which there are three fixed values a, b, c such that the roots of the equations*

$$f(z) = a, \; f(z) = b, \; f(z) = c$$

have multiplicity divisible by h, k, ℓ, respectively, for each $f \in \mathcal{F}$. If

$$\frac{1}{h} + \frac{1}{k} + \frac{1}{\ell} < 1,$$

then $\mathcal{F}$ is normal in Ω.

We defer proof of this theorem since a stronger version will be proved in Chapter 4 from two different points of view.

Vitali-Porter Theorem. It is evident from the proof of the Vitali-Porter Theorem that a normal family $\{f_n\}$ of analytic functions such that $\lim_{n\to\infty} f_n(z)$ exists at a sequence of points having a limit in Ω must in fact converge uniformly on compact subsets of Ω. The natural analogue for meromorphic functions, since normality is equivalent to equicontinuity, is

Theorem 3.2.3 *Let $\{f_n\}$ be a sequence belonging to a spherically equicontinuous family of meromorphic functions such that $\{f_n(z)\}$ converges spherically on a point set E having a limit in Ω. Then $\{f_n\}$ converges spherically uniformly on compact subsets of Ω.*

The proof is a straightforward adaptation of that for the Vitali-Porter Theorem, by invoking the Identity Theorem for meromorphic functions.

3.3   Marty's Theorem

In contradistinction to Montel's theorem, where local boundedness was the key ingredient, the normality of a family of meromorphic functions is characterized by a condition in which the spherical derivative is locally bounded, due to Marty [1931]; cf. Hayman [1964], pp. 158–160 for another proof, also Hua and Chen [preprint] for a generalization reminiscent of the Five Islands Theorem (§1.9).

Marty's Theorem *A family $\mathcal{F}$ of meromorphic functions on a domain Ω is normal if and only if for each compact subset $K \subseteq \Omega$, there exists a constant $C = C(K)$ such that the spherical derivative*

$$f^{\#}(z) = \frac{|f'(z)|}{1 + |f(z)|^2} \leq C, \qquad z \in K, \; f \in \mathcal{F},$$

that is, $f^{\#}$ is locally bounded.

Proof. Suppose that $f^{\#}(z)$ is uniformly bounded on compact subsets. Take an arbitrary $z_0 \in \Omega$ and a closed disk $K(z_0; r) \subseteq \Omega$. For any $z \in K(z_0; r)$, let γ be the straight-line path joining z_0 to z in Ω, so that

$$\chi\big(f(z_0), f(z)\big) \leq \int_{\gamma} f^{\#}(\zeta)|d\zeta|.$$

Therefore, for some constant $C = C(z_0), \chi\big(f(z_0), f(z)\big) \leq C|z_0 - z|, f \in \mathcal{F}$, and we conclude that $\mathcal{F}$ is spherically equicontinuous, hence normal in Ω.

Conversely, suppose that $\mathcal{F}$ is normal, but there is a compact set $K \subseteq \Omega$, a sequence of points $\{z_n\} \subset K$ and a sequence of functions $\{f_n\} \subseteq \mathcal{F}$ such that $f_n^{\#}(z_n) \to \infty$ as $n \to \infty$. Firstly we obtain a subsequence $\{f_{n_k}\}$ that converges spherically uniformly on K to a limit f. Centred about each $z_0 \in K$ there is a closed disk $K(z_0; r) \subseteq \Omega$ in which either $f_{n_k} \to f$ or $\frac{1}{f_{n_k}} \to \frac{1}{f}$ uniformly as $k \to \infty$ (Theorem 3.1.3).

In the former case, f is analytic and bounded in $K(z_0; r)$ (that is, on a slightly larger open disk), whence f_{n_k} is analytic in $K(z_0; r)$ for k sufficiently large. Then $f_{n_k}^{\#} \to f^{\#}$ by the Weierstrass Theorem, uniformly on $K(z_0; r)$. Since $f^{\#}(z)$ is bounded on $K(z_0; r)$, the same is true for the $f_{n_k}^{\#}$'s.

In the event $\frac{1}{f_{n_k}} \to \frac{1}{f}$, replace f_{n_k} and f by $\frac{1}{f_{n_k}}$ and $\frac{1}{f}$, respectively, in the above argument, and since $g^{\#}(z) = \left(\frac{1}{g(z)}\right)^{\#}$, we obtain the same conclusion. As K is compact, it can be covered by a finite number of disks in each of which the $f_{n_k}^{\#}$ are bounded, so that the $f_{n_k}^{\#}$ are bounded on K. This contradiction concludes the proof of the theorem.

The Marty criterion is one of the most widely used for determining the normality of a family of meromorphic functions. However, given a family of meromorphic functions $\{f\}$ such that, say,

$$|f'| \leq e^{|f|},$$

we see that the Marty criterion is insufficient to establish normality, and a stronger version is required. The following has been furnished by Royden [1985].

Theorem 3.3.1 *Let $\mathcal{F}$ be a family of meromorphic functions in a domain Ω such that for each compact subset K of Ω there is a monotone increasing function h_K such that*

$$|f'(z)| \leq h_K\big(|f(z)|\big), \qquad f \in \mathcal{F}, \ z \in K.$$

Then $\mathcal{F}$ is normal in Ω.

Proof. It suffices to consider the case when Ω is an open disk and $|f'(z)| \leq h\big(|f(z)|\big)$ there. Moreover, as this inequality certainly holds for all larger h, we can assume that h is continuous and that $h(t) > 1 + t^2$. If γ is a differentiable arc or curve on the Riemann sphere Σ define the *h-length* of γ by

$$L_h(\gamma) = \int_\gamma \frac{|dw|}{h(|w|)}.$$

Then $L_h(\gamma)$ is less than the spherical length of γ, $L(\gamma)$ (§1.2).

Define $\rho : \Sigma \times \Sigma \to \mathbb{R}$ by

$$\rho(w_1, w_2) = \inf\{L_h(\gamma)\},$$

where the infimum is taken over all γ joining w_1 with w_2. Clearly ρ is a metric, and $\rho(w_1, w_2) \leq \sigma(w_1, w_2)$, the spherical metric on Σ. Therefore, ρ and σ are uniformly equivalent owing to the compactness of Σ in σ.

As in Marty's theorem, letting Γ be the straight-line path joining z_1 with z_2, we have for each $f \in \mathcal{F}$

$$\rho\big(f(z_1), f(z_2)\big) \leq \int_\Gamma \frac{|f'(z)||dz|}{h(|f(z)|)} \leq \int_\Gamma |dz| = |z_1 - z_2|,$$

invoking the criterion of the hypothesis. Thus, the family $\mathcal{F}$ is equicontinuous with respect to the metric ρ and, in view of Theorem 3.2.1, $\mathcal{F}$ is normal with respect to ρ. The normality of $\mathcal{F}$ with respect to the chordal metric χ now follows due to the uniform equivalence of ρ and χ.

It is now evident that the family of meromorphic functions $\{f\}$ such that $|f'| \leq e^{|f|}$ is normal. An improvement of the Royden criterion has been given by Chen and Gu [preprint], cf. also Hua [preprint].

As in the case of analytic functions, we say that a family of meromorphic functions $\mathcal{F}$ is *normal at a point z*, if $\mathcal{F}$ is normal in some neighbourhood of z. As for analytic functions, normality is a local property.

Theorem 3.3.2 *A family of meromorphic functions is normal in a domain Ω if and only if it is normal at each of its points.*

The necessity is obvious, and the sufficiency follows readily from Marty's theorem via a compactness argument.

In accordance with the analytic case, we define a family of meromorphic functions $\mathcal{F}$ to be *normal at ∞* if the corresponding family $\mathcal{G} = \{g : g(z) = f\left(\frac{1}{z}\right)\}$ is normal at $z = 0$. Moreover, if $\Omega \subseteq \mathbb{C}$ and Ω contains the point at infinity, define $\mathcal{F}$ to be *normal in Ω* if $\mathcal{F}$ is normal in $\Omega - \{\infty\}$ in the usual sense, as well as normal at ∞. This is equivalent to saying that: *A family of meromorphic functions $\mathcal{F}$ is normal in a domain Ω on the Riemann sphere Σ if and only if every sequence $\{f_n\} \subseteq \mathcal{F}$ contains a subsequence that converges spherically uniformly on compact subsets of Ω.* It also follows that many results on normality, such as the FNT (§3.2), Theorem 3.2.1, etc., have immediate extensions to a domain $\Omega \subseteq \Sigma$.

An interesting characterization of normal families now arises which is somewhat reminiscent of Theorem 3.1.3. Observe that the chordal distance between the circles $|w| = 1$ and $|w| = 2$ is given by $\chi(1,2) = \frac{1}{\sqrt{10}}$, which is the chordal distance between the circles $|w| = 1$ and $|w| = \frac{1}{2}$. Then for any two points w', w'' that satisfy either $|w'| \leq 1$ and $|w''| \geq 2$, or $|w'| \geq 1$ and $|w''| \leq \frac{1}{2}$, it must be true that

$$\chi(w', w'') \geq \frac{1}{\sqrt{10}}.$$

Suppose now that $\mathcal{F}$ is a normal family of meromorphic functions in Ω. In view of Theorem 3.2.1, about each point $z_0 \in \Omega$, there is a neighbourhood $N(z_0) \subseteq \Omega$ in which

$$\chi\big(f(z), f(z_0)\big) < \frac{1}{\sqrt{10}},$$

for all $f \in \mathcal{F}$. If, say, $|f(z_0)| \leq 1$, then by the preceding remarks $|f(z)| < 2$, for $z \in N(z_0)$, $f \in \mathcal{F}$, whereas if $|f(z_0)| > 1$, then $|f(z)| > \frac{1}{2}$. Therefore, if $z \in N(z_0)$, either

$$|f(z)| < 2 \quad \text{or} \quad \frac{1}{|f(z)|} < 2, \qquad f \in \mathcal{F}.$$

Conversely, if this latter condition holds in $N(z_0)$, then $\mathcal{F}$ may be expressed as the union of the two families

$$\mathcal{G}_0 = \{f \in \mathcal{F} : |f(z)| < 2, \ z \in N(z_0)\},$$

$$\mathcal{H}_0 = \{f \in \mathcal{F} : |f(z)| > \tfrac{1}{2}, \ z \in N(z_0)\}.$$

It follows that $\mathcal{G}_0$ and H_0 are normal at z_0, so that the union $\mathcal{F}$ is normal at z_0 and hence normal in Ω by Theorem 3.3.2. We have established

Theorem 3.3.3 *If a family of meromorphic functions $\mathcal{F}$ is normal in a domain Ω, then about each point $z_0 \in \Omega$, there is a neighbourhood $N(z_0) \subseteq \Omega$ such that either*

$$|f(z)| < 2 \quad or \quad \frac{1}{|f(z)|} < 2$$

holds for all $z \in N(z_0)$, $f \in \mathcal{F}$, and conversely.

The partitioning of the family $\mathcal{F}$ into two normal families works equally well if the constant 2 in the above is replace by an arbitrary positive constant.

Corollary 3.3.4 *Let $\mathcal{F}$ be a family of meromorphic functions in Ω such that for each $z_0 \in \Omega$, there is a neighbourhood $N(z_0) \subseteq \Omega$ and a constant $M(z_0)$ such that either*

$$|f(z)| < M(z_0) \quad or \quad \frac{1}{|f(z)|} < M(z_0), \qquad z \in N(z_0), f \in \mathcal{F}.$$

Then $\mathcal{F}$ is normal in Ω.

3.4   Compactness

From the analytic case, we say that a family $\mathcal{F}$ of meromorphic functions is *compact* if the limit of every normally convergent sequence belongs to $\mathcal{F}$. In this regard we maintain (cf. Carathéodory [1958, Vol. I], p. 183).

Theorem 3.4.1 *Let $\mathcal{F}$ be a normal family of meromorphic functions on a domain Ω. Then $\mathcal{F}$ can be extended to a normal and compact family by adding to $\mathcal{F}$ all limits of normally convergent sequences in $\mathcal{F}$.* (Here, with an *abuse de langue*, we treat $f \equiv \infty$ as a meromorphic function.)

Proof. Denote by $\mathcal{G}$, the extended family with the limits adjoined. We may assume every $g \in \mathcal{G}$ is a limit of functions in $\mathcal{F}$ by admitting sequences in which the same function is repeated infinitely often. Then every $g \in \mathcal{G}$ is a meromorphic function in Ω.

In order to show $\mathcal{G}$ is normal in Ω, consider an arbitrary sequence $\{g_n\} \subseteq \mathcal{G}$, and an exhaustion $\{K_n\} \subseteq \Omega$. Then for $n = 1, 2, 3, \dots$, there is a sequence $\{f_k^{(n)}\} \subseteq \mathcal{F}$ which converges normally to g_n as $k \to \infty$. Hence for each n there is an integer $k(n)$ sufficiently large satisfying

$$\chi\Big(f_{k(n)}^{(n)}(z), g_n(z)\Big) < \frac{1}{n}, \qquad z \in K_n. \tag{3.2}$$

Since $\mathcal{F}$ is normal, the sequence $\{f_{k(n)}^{(n)}\}$ has a subsequence $\left\{f_{k(n_j)}^{(n_j)}\right\}$ which converges spherically uniformly on compact subsets to a meromorphic function g. By (3.2) we deduce that $\{g_{n_j}\}$ converges spherically uniformly to g on compact subsets of Ω, i.e., $\mathcal{G}$ is a normal family. Moreover, since any limit function g of functions $\{g_n\} \subseteq \mathcal{G}$ is also a limit of a sequence of functions in $\mathcal{F}$, we conclude that $\mathcal{G}$ is compact.

Our proof is obviously a special case of what can be done in a more general metric space setting.

If $\mathcal{F}$ is a family of analytic functions that does not admit the limit function $\equiv \infty$, say if $|f(z_0)| \leq M$ for all $f \in \mathcal{F}$, then $\mathcal{F}$ may be compactified in the same fashion.

3.5   Poles of a Normal Family

For a normal family of analytic functions $\mathcal{F}$ that has no constant limit function a, the number of roots of $f(z) - a$ is locally uniformly bounded for all $f \in \mathcal{F}$ (Theorem 2.5.1). For meromorphic functions we have a similar condition on the number of poles (Montel [1916]).

Theorem 3.5.1 *Let $\mathcal{F}$ be a normal family of meromorphic functions in a domain Ω such that for some $z_0 \in \Omega$, $|f(z_0)| \leq M$, for all $f \in \mathcal{F}$. Then* *the number of poles of each $f \in \mathcal{F}$ is uniformly bounded on each compact subset of Ω.*

The proof follows along the same general lines as Theorem 2.5.1. Assume that the number of poles on a compact subset $K \subseteq \Omega$ is not bounded. Then for $n = 1, 2, 3, \ldots$, there are functions $f_n \in \Omega$ which have at least n poles in K. The hypotheses guarantee a subsequence $\{f_{n_k}\}$ which converges spherically uniformly on K to a meromorphic function $f(z) \not\equiv \infty$. The function $f(z)$ can have only a finite number of poles in K, say $z_1, z_2, \ldots, z_p$, with orders $\alpha_1, \alpha_2, \ldots, \alpha_p$, respectively. For $i = 1, \ldots, p$, consider a set of mutually disjoint disks $D(z_i; r_i)$, which we may assume are subsets of K and in which $\frac{1}{f_{n_k}} \to \frac{1}{f}$ uniformly. As the functions $\frac{1}{f_{n_k}(z)}$ and $\frac{1}{f(z)}$ are analytic in each $D(z_i; r_i)$, they have the same number of zeros there for k sufficiently large. Then for such k, $f_{n_k}(z)$ has poles of total order α_i in $D(z_i; r_i)$, and hence $\alpha = \alpha_1 + \ldots + \alpha_p$ poles in $\cup_{i=1}^{p} D(z_i; r_i)$.

In the complement $E = K - \cup_{i=1}^{p} D(z_i; r_i)$, $f(z)$ is analytic and bounded. Then $f_{n_k} \to f$ uniformly on E, implying $\{f_{n_k}\}$ for large k has no further poles in E, and this contradicts the assumption that the number of poles in K increases without bound.

Corollary 3.5.2 *Let $\mathcal{F}$ be a normal family of meromorphic functions in Ω which satisfy, at some point $z_0 \in \Omega$,*

$$|f(z_0) - a| > r, \qquad f \in \mathcal{F},$$

for some $r > 0$, $a \in \mathbb{C}$. Then the number of roots of the equation $f(z) - a$ is uniformly bounded on each compact subset of Ω.

Proof. The family

$$\mathcal{G} = \left\{ g(z) = \frac{1}{f(z) - a} : f \in \mathcal{F} \right\}$$

consists of meromorphic functions satisfying $|g(z_0)| < \frac{1}{r}$. The result then follows from the theorem.

If a normal family is bounded at a certain point, something further can be said about the proximity of any pole of a function in the family.

Theorem 3.5.3 *Let $\mathcal{F}$ be a normal family of meromorphic functions in a domain Ω such that $|f(z_0)| \leq M$ for each $f \in \mathcal{F}$. Then for some $r > 0$, there is a disk $D(z_0; r)$ in which there is no pole of any $f \in \mathcal{F}$.*

Proof. Suppose this were not the case. Then there is a sequence of functions $\{f_n\} \subseteq \mathcal{F}$ having poles $\{z_n\}$, such that $z_n \to z_0$ as $n \to \infty$. We extract a subsequence $\{f_{n_k}\}$ converging spherically uniformly on compact subsets to a meromorphic function f. Then $|f(z_0)| \leq M$, and indeed there is a disk $D(z_0; r)$ in which $|f(z)| \le M'$, say. Choose k large enough so that $z_{n_k} \in D(z_0; r)$ and $|f_{n_k}(z) - f(z)| < 1$ for all $z \in D(z_0; r)$. Then

$$|f_{n_k}(z)| < M' + 1, \qquad z \in D(z_0; r),$$

and this contradiction establishes the theorem.

Corollary 3.5.4 *Let $\mathcal{F}$ be a family of meromorphic functions in Ω such that*

$$|f(z_0) - a| > r, \qquad f \in \mathcal{F},$$

for some $r > 0$, $z_0 \in \Omega$, $a \in \mathbb{C}$. Then for each $f \in \mathcal{F}$, the family of equations $f(z) - a = 0$ has no roots in some neighbourhood of z_0.

Indeed, the functions

$$g(z) = \frac{1}{f(z) - a}, \qquad f \in \mathcal{F},$$

satisfy the conditions of the theorem.

As an application of the preceding theorem, let us establish a meromorphic version of Landau's theorem (§2.8). Toward this end, consider the family $\mathcal{F}$ of meromorphic functions in $|z| < R < \infty$ such that any $f \in \mathcal{F}$ has the Taylor expansion

$$f(z) = a_0 + a_1 z + \ldots, \qquad a_1 \ne 0,$$

in some neighbourhood of the origin. Suppose that $\mathcal{F}$ happens to be normal in $|z| < R$ by virtue of some property $\mathcal{P}$ which is invariant under homothetic transformations of the domain. Define

$$\widetilde{\mathcal{F}} = \{\tilde{f} : \tilde{f}(z) = f(Rz), \ f \in \mathcal{F}\}, \qquad |z| < 1.$$

Then $\widetilde{\mathcal{F}}$ is normal in $|z| < 1$, with $\tilde{f}(0) = f(0) = a_0$, $\tilde{f} \in \widetilde{\mathcal{F}}$. In view of Theorem 3.5.3, each $\tilde{f}$ is analytic in a disk $|z| < \rho = \rho(a_0, \mathcal{P})$. The important feature to note is that ρ is independent of R, as well as f. We conclude that each $f \in \mathcal{F}$ is analytic in $|z| < r = \rho R$.

Applying Corollary 2.2.4 gives

$$|f(z)| \le M(a_0, \mathcal{P}), \qquad |z| \le \frac{r}{2}, f \in \mathcal{F},$$

and by the Cauchy formula

$$\begin{aligned} |a_1| &\le \frac{M(a_0, \mathcal{P})}{\frac{r}{2}} \\ &= \frac{2M(a_0, \mathcal{P})}{\rho R}. \end{aligned}$$

Hence

$$R \leq \frac{2M(a_0, \mathcal{P})}{\rho(a_0, \mathcal{P})|a_1|} = M(a_0, a_1, \mathcal{P}).$$

This leaves only the matter of the normality of $\mathcal{F}$. One possibility is for $\mathcal{F}$ to omit three distinct values. A more general condition implying normality is embraced in the following hypotheses (cf. Theorems 4.1.3 and 4.1.4).

Theorem 3.5.5 *Let $\mathcal{F}$ be a family of meromorphic functions in the disk $|z| < R$ such that each function $f \in \mathcal{F}$ has the expansion*

$$f(z) = a_0 + a_1 z + \ldots, \qquad a_1 \neq 0,$$

in a neighbourhood of the origin. Suppose that all the roots of the equation $f(z) - \alpha_\nu = 0$, $\alpha_\nu \in \widehat{\mathbb{C}}$ have multiplicity $\geq m_\nu$ (≥ 2), $\nu = 1, \ldots, q$, and

$$\sum_{\nu=1}^{q} \left(1 - \frac{1}{m_\nu} \right) > 2.$$

Then

$$R \leq M(a_0, a_1, m_1 \ldots, m_q).$$

If the values $0, 1, \infty$ are omitted by $\mathcal{F}$, then the corresponding m_ν's are taken to be ∞, and the usual Landau Theorem obtains. For Schottky type theorems pertaining to meromorphic functions, see Valiron [1929], pp. 39–40.

3.6   Invariant Normal Families

Marty's theorem can be rendered into a more precise form when applied to a family which is conformally invariant. This notion, along with the initial development in this section, is due to Hayman [1955, 1964].

Definition 3.6.1 *A family $\mathcal{F}$ of meromorphic functions defined on a simply connected domain Ω is* **(conformally) invariant** *if, whenever $f \in \mathcal{F}$, then $f \circ t \in \mathcal{F}$, where t is any one-to-one conformal mapping of Ω onto Ω.*

Invariant families are intimately related to the notion of normal functions, which are taken up in Chapter 5, although the notions were developed independently.

The normality of such families is readily determined for $\Omega = U$.

Theorem 3.6.2 *Let $\mathcal{F}$ be an invariant family of meromorphic functions in U. Then $\mathcal{F}$ is normal in U if and only if for some constant M,*

$$f^{\#}(0) = \frac{|f'(0)|}{1 + |f(0)|^2} \leq M \tag{3.3}$$

for all $f \in \mathcal{F}$.

Proof. Certainly if $\mathcal{F}$ is normal, then (3.3) holds by Marty's theorem.

On the other hand, if (3.3) holds, then applying it to $g(z)$ $= f\left(\dfrac{z+a}{1+\bar{a}z}\right) \in \mathcal{F}$ gives

$$\frac{(1-|a|^2)|f'(a)|}{1+|f(a)|^2} \leq M, \quad f \in \mathcal{F}.$$

For a belonging to a compact subset of U, the conclusion again follows from Marty's theorem.

If $w = f(z)$ is meromorphic in U, recall from §1.2:

$$
\begin{aligned}
L(r) &= \text{length of the image of } |z| = r \text{ on the Riemann sphere} \\
&= \int_0^{2\pi} \frac{|f'(re^{i\theta})|r\,d\theta}{1+|f(re^{i\theta})|^2}; \\
S(r) &= \frac{1}{\pi} \cdot \text{area of the image of } |z| < r \text{ on the Riemann sphere with} \\
&\quad \text{due regard to multiplicity} \\
&= \frac{1}{\pi} \int_0^{2\pi} \int_0^r \frac{|f'(\rho e^{i\theta})|^2 \rho\,d\rho\,d\theta}{\left(1+|f(\rho e^{i\theta})|^2\right)^2}.
\end{aligned}
$$

The following theorem due to Dufresnoy [1941] will be required. The proof in Hayman [1964], p. 152, is based on the isoperimetric inequality on the sphere.

Dufresnoy's Theorem *If $f(z)$ is meromorphic in $|z| \leq r_0$ and $S(r_0)$ $= S(r_0, f) < 1$, then*

$$[f^{\#}(0)]^2 \leq \frac{1}{r_0^2} \frac{S(r_0)}{1-S(r_0)}.$$

We can now establish the following sufficient condition for normality from the uniform boundedness of the spherical areas. The analogous result of Example 2.2.7 for analytic functions defined in U required the uniform boundedness of the Riemann surface areas of the images. In that particular case, we see that the family would also be invariant since the Riemann surface area of a Dirichlet-finite function and its transpose are the same.

Theorem 3.6.3 *Let $0 < K < 1$ and $\mathcal{F}_K$ be the class of meromorphic functions in U such that $S(r) = S(r, f) \leq K$ for $0 < r < 1$. Then $\mathcal{F}_K$ is an invariant normal family.*

Proof. By virtue of the Dufresnoy Theorem,

$$f^{\#}(0) \leq \frac{1}{r}\sqrt{\frac{K}{1-K}},$$

and letting $r \to 1$,

$$f^{\#}(0) \leq \sqrt{\frac{K}{1-K}}.$$

Moreover, if $f(z) \in \mathcal{F}_K$, then $g(z) = f\left(e^{i\gamma}\dfrac{z+a}{1+\bar{a}z}\right) \in \mathcal{F}_K$, for γ real, $|a| <$ 1, since $f(U)$ and $g(U)$ have the same images on the sphere, the area of which is not greater than πK. Therefore, $\mathcal{F}_K$ is invariant, and by Theorem 3.6.2, $\mathcal{F}_K$ is normal.

The normality of $\mathcal{F}_K$ was already mentioned in Montel [1934].
We remark that by Example 2.1.3 the family

$$\{f_n(z) = nz : n = 1, 2, 3, \ldots\}$$

is not normal in U. On the other hand, $S(r, f_n) < 1$, $0 < r < 1$, so that the class $\mathcal{F}_1$ is not normal in U.

At this juncture it is appropriate to investigate those meromorphic functions in U for which the length-area relation

$$S(r) < hL(r)$$

holds for some positive constant h. Such functions arise in the context of the Ahlfors Five Islands Theorem (cf. §1.9 and Corollary 3.6.6).

Lemma 3.6.4 *If $f(z)$ is meromorphic in U such that*

$$S(r, f) = S(r) < h_1 L(r), \qquad 0 < r < 1, \tag{3.4}$$

for some constant $h_1 > 0$, then

$$f^{\#}(0) < h_2, \tag{3.5}$$

where h_2 depends only on h_1.

Proof. For $0 < r_0 < r < 1$, we apply the Schwarz inequality to

$$L(r) = \int_0^{2\pi} \frac{|f'(re^{i\theta})|r\,d\theta}{1 + |f(re^{i\theta})|^2}.$$

This gives

$$L^2(r) \leq \int_0^{2\pi} r\,d\theta \int_0^{2\pi} \frac{|f'(re^{i\theta})|^2 r\,d\theta}{(1 + |f(re^{i\theta})|^2)^2} = 2\pi^2 r S'(r),$$

so that

$$S^2(r) \leq 2\pi^2 h_1^2 r S'(r)$$

from our assumption (3.4). Hence

$$\frac{1}{S(r_0)} - \frac{1}{S(r)} = \int_{r_0}^{r} \frac{S'(t)}{S^2(t)} \, dt \geq \frac{1}{2\pi^2 h_1^2} \int_{r_0}^{r} \frac{dt}{t} = \frac{\log \frac{r}{r_0}}{2\pi^2 h_1^2},$$

and we deduce that

$$S(r_0) \leq \frac{2\pi^2 h_1^2}{\log \frac{r}{r_0}}.$$

Letting $r \to 1$,

$$S(r_0) \leq \frac{2\pi^2 h_1^2}{\log \frac{1}{r_0}}, \qquad 0 < r_0 < 1.$$

Finally, choose r_0 satisfying

$$\frac{2\pi^2 h_1^2}{\log \frac{1}{r_0}} = \frac{1}{2},$$

i.e., $r_0 = e^{-4\pi^2 h_1^2}$. Then Dufresnoy's theorem implies

$$[f^{\#}(0)]^2 \leq \frac{1}{r_0^2} \frac{S(r_0)}{1 - S(r_0)} \leq \frac{1}{r_0^2} = e^{8\pi^2 h_1^2},$$

so that (3.5) holds.

This leads to another invariant normal family.

Theorem 3.6.5 *Suppose that $f(z)$ is meromorphic in U and let L be the length of the image of the circumference on Σ and πS be the area of the image on Σ of some disk whose closure lies in U. Define $\mathcal{F}_h$ as the class of all functions for which*

$$S \leq hL$$

for all such disks, where $h > 0$ is constant. Then $\mathcal{F}_h$ is an invariant normal family.

Proof. Let D be a disk with $\overline{D} \subseteq U$. Under the mapping

$$t(z) = e^{i\gamma} \frac{z + a}{1 + \bar{a}z},$$

D is mapped onto a disk Δ with $\bar{\Delta} \subseteq U$. Given $f \in \mathcal{F}_h$, the function

$$g(z) = f\big(t(z)\big)$$

then satisfies $g(D) = f(\Delta)$. Thus, $g \in \mathcal{F}_h$ and $\mathcal{F}_h$ is invariant. Furthermore, by Lemma 3.6.4 we have the requisite condition for the normality of $\mathcal{F}_h$ (Theorem 3.6.2).

In view of the Ahlfors Five Islands Theorem (§1.9) we have

Corollary 3.6.6 *Suppose that $\mathcal{F}$ comprises the family of meromorphic functions $w = f(z)$ which map no subdomain of U one-to-one conformally onto any of five given Jordan domains (with s.a. boundaries) whose closures are disjoint on Σ. Then $\mathcal{F}$ is normal and invariant.*

In fact, by §1.9, there is a constant $h > 0$ independent of f for which

$$S(r) < hL(r), \qquad 0 < r < 1,$$

and the result follows from the theorem.

In the present context we can deduce another proof of the FNT (§2.7). For, a function f belonging to the family

$$\mathcal{F} = \{f \text{ analytic in } U : f(z) \neq 0, 1\}$$

will possess no islands over any domains which contain the points $0, 1, \infty$ on the sphere. Then from the Second Fundamental Theorem of Ahlfors, we can assert that

$$S(r) < hL(r), \qquad 0 < r < 1,$$

for some absolute constant h. From Lemma 3.6.4, $f^{\#}(0) \leq C$, for all $f \in \mathcal{F}$, and since $\mathcal{F}$ is an invariant family, it is normal by Theorem 3.6.2, and the proof is complete.

Given an invariant normal family, we now show how this constrains the growth of the Ahlfors-Shimizu characteristic.

Theorem 3.6.7 *If $\mathcal{F}$ is an invariant normal family, then*

$$T_0(r, f) \leq C \log \frac{1}{1 - r^2}, \qquad 0 < r < 1, \ f \in \mathcal{F},$$

where $C > 0$ is constant.

Proof. Since $f \in \mathcal{F}$ implies $f^{\#}(0) \leq C$, for some $C > 0$, we have for $f\left(\dfrac{z + z_0}{1 + \bar{z}_0 z}\right) \in \mathcal{F}$, $z_0 = re^{i\theta}$, the estimate

$$[f^{\#}(z)]^2 \leq \frac{C^2}{(1 - r^2)^2}, \qquad |z| = r.$$

This gives, upon integrating,

$$\begin{aligned}
S(r, f) &= \frac{1}{\pi} \int_0^{2\pi} \int_0^r [f^{\#}(\rho e^{i\theta})]^2 \rho \, d\rho \, d\theta \leq 2 C^2 \int_0^r \frac{\rho \, d\rho}{(1 - \rho^2)^2} \\
&= C^2 \frac{r^2}{1 - r^2}.
\end{aligned}$$

Therefore,

$$T_0(r, f) = \int_0^r \frac{S(t)}{t}\, dt \leq \frac{C^2}{2} \log \frac{1}{1 - r^2},$$

as desired.

Working locally, we can determine a necessary and sufficient condition for normality in terms of $T_0(r, f)$ by similar means (Drasin [1988]).

Theorem 3.6.8 *If $\mathcal{F}$ is a family of meromorphic functions in U, then $\mathcal{F}$ is normal in a neighbourhood of the origin if and only if there is a constant $C > 0$ such that*

$$T_0(r, f) \leq C, \qquad f \in \mathcal{F},\ 0 < r < R,$$

for some $R > 0$.

Proof. Suppose $\mathcal{F}$ is normal in some neighbourhood of the origin. Then by Marty's criterion,

$$f^{\#}(z) \leq M$$

for $|z| \leq R$, $R > 0$. Therefore,

$$S(r) = S(r, f) = \frac{1}{\pi} \int_0^{2\pi} \int_0^r [f^{\#}(\rho e^{i\theta})]^2 \rho\, d\rho\, d\theta \leq M^2 r^2, \quad 0 \leq r < R,\ f \in \mathcal{F},$$

and so

$$T_0(r, f) = \int_0^r \frac{S(t)}{t}\, dt \leq \frac{M^2 r^2}{2} < C, \qquad 0 < r < R,\ f \in \mathcal{F}.$$

The proof of the converse was suggested to the author by Professor Hayman. If $T_0(r, f) \leq C$, $0 < r < R$, then for $0 < r_0 < r$, r fixed, we have, as in (1.8),

$$C \geq \int_0^r \frac{S(t)}{t}\, dt > S(r_0) \int_{r_0}^r \frac{dt}{t} = S(r_0) \log \frac{r}{r_0}.$$

It follows that

$$S(r_0) < \frac{C}{\log \frac{r}{r_0}} < \frac{1}{2},$$

for r_0 sufficiently small, and the normality at the origin is a consequence of a local version of Theorem 3.6.3.

Considering the same question for the Nevanlinna characteristic, we maintain (Schwick [1989], Drasin [1969], Lemma 4).

Theorem 3.6.9 *Let $\mathcal{F}$ be a family of meromorphic functions in U such that for each $f \in \mathcal{F}$, $f(0) \neq \infty$, and*

$$T(r_0, f) < C < \infty$$

for some $0 < r_0 < 1$. Then $\mathcal{F}$ is normal in a neighbourhood of the origin.

Proof. Given $f \in \mathcal{F}$, suppose that b is a pole of f with $|b| < r_0$. Then

$$\log \frac{r_0}{|b|} \leq N(r_0, f) \leq T(r_0, f) < C,$$

that is,

$$|b| > r_0 e^{-C} = r_1 > 0.$$

Hence every $f \in \mathcal{F}$ is analytic in $|z| \leq r_1$. Moreover, by Theorem 1.8.2

$$\log^+ M\left(\frac{r_1}{2}, f\right) \leq \frac{r_1 + r_1/2}{r_1 - r_1/2} T(r_1, f) < 3C,$$

and the conclusion follows by Montel's theorem (§2.2).

In contrast with the Ahlfors-Shimizu characteristic, the converse of Theorem 3.6.9 is not valid. Indeed, fix $a \neq 0$, $|a| < \frac{1}{4}$, and set

$$f_n(z) = (z - a)^{-n}, \qquad n = 1, 2, 3, \ldots .$$

Then for $|z| < r_0 < |a|$, $|z - a|^{-n} > 2^n$, so that $f_n \to \infty$ normally in $|z| < r_0$. However, for $0 < r < r_0$

$$T(r, f_n) = m(r, f_n) = -n \cdot \frac{1}{2\pi} \int_0^{2\pi} \log |re^{i\theta} - a| \, d\theta = -n \log |a|$$

by Jensen's formula (1.5), and $T(r, f_n) \to \infty$ as $n \to \infty$.

Returning to invariant normal families, our intention next is to demonstrate (for analytic functions) a generalization of the Schottky and Landau Theorems which is due to Hayman [1955, 1964].

Before preceeding with the main result, let us first consider a suggestive example; specifically the family

$$\mathcal{F}_\alpha = \{f(z) \text{ analytic in } U : |f(z)| > \alpha > 0\}.$$

Then $\mathcal{F}_\alpha$ is an invariant normal family in U. If we set

$$F(z) = \frac{f(z)}{\alpha}, \qquad |z| < 1,$$

the function $h(z) = \log |F(z)|$ is therefore positive harmonic. In view of the Herglotz representation for $h(z)$, namely

$$u(z) = \frac{1}{2\pi} \int_0^{2\pi} \frac{1 - |z|^2}{|\zeta - z|^2} \, d\mu(\phi), \qquad \zeta = e^{i\phi}, \ |z| = r < 1,$$

where $\mu(\phi)$ is a nondecreasing function on $[0, 2\pi]$ (cf. §5.4), it follows that

$$\log |F(z)| \leq \frac{1+r}{1-r} \log |F(0)|.$$

In other words,

$$\log|f(z)| \le \frac{1+r}{1-r}\log|f(0)| - \frac{2r\log\alpha}{1-r}.$$

Setting $C = -\log\alpha$ allows us to write

$$\log M(r,f) \le \frac{1+r}{1-r}\log^+|f(0)| + \frac{2Cr}{1-r},$$

and it is this inequality it turns out, which is valid for an arbitrary invariant normal family of analytic functions in the unit disk.

Theorem 3.6.10 *Let $\mathcal{F}$ be an invariant normal family of analytic functions in U. Then for $f \in \mathcal{F}$,*

$$f(z) = a_0 + a_1 z + \dots \qquad |z| < 1,$$

there is a constant C depending only on $\mathcal{F}$ such that

$$|a_1| \le 2\mu(\log\mu + C)$$

and

$$M(r,f) \le \mu^{(1+r)/(1-r)} e^{2Cr/(1-r)},$$

where $\mu = \max(1, |a_0|)$.

In order to prove the theorem, two lemmas are required.

Lemma 3.6.11 *If $\mathcal{F}$ is as in the theorem, then there are positive constants A and r_0, such that for any $f \in \mathcal{F}$ with $|f(z_1)| \le 1$, $|f(z_2)| \ge e^A$, $z_1, z_2 \in U$, the inequality*

$$\left|\frac{z_1 - z_2}{1 - \bar{z}_1 z_2}\right| \ge r_0$$

holds.

Proof. If this were not the case, then one could find sequences $\{f_n\} \subseteq \mathcal{F}$, $\{z_n\}, \{z'_n\}$ in U that satisfy

$$|f_n(z_n)| \le 1, \quad |f_n(z'_n)| \to \infty, \quad \text{and} \quad \left|\frac{z_n - z'_n}{1 - \bar{z}_n z'_n}\right| \to 0,$$

as $n \to \infty$. Setting

$$g_n(z) = f_n\left(\frac{z + z_n}{1 + \bar{z}_n z}\right),$$

then $g_n \in \mathcal{F}$ and $|g_n(0)| \le 1$. Furthermore,

$$z'_n = \frac{z_n + \zeta'_n}{1 + \bar{z}_n \zeta'_n},$$

for some $\zeta_n' \in U$, so that $|g_n(\zeta_n')| \to \infty$ and

$$|\zeta_n'| = \left| \frac{z_n - z_n'}{1 - \bar{z}_n z_n'} \right| \to 0.$$

Since $\mathcal{F}$ is normal in U, $\{g_n\}$ has a subsequence which we again denote by $\{g_n\}$, which converges normally in U to an analytic function g or to ∞. In the former case, $|g(0)| \le 1$, whereas $|g_n(\zeta_n')| \to \infty$, leading to a contradiction. Nor can the limit be ∞ since $|g_n(0)| \le 1$, completing the proof.

Lemma 3.6.12 *Let*

$$F(z) = b_0 + b_1 z + \ldots$$

be analytic in U and satisfy $|F(z)| > 1$ there. Then

$$|b_1| \le 2|b_0| \log |b_0|$$

and

$$|F(z_1)| \ge |F(z_2)|^{(1-t)/(1+t)},$$

for any $z_1, z_2 \in U$, where

$$t = \left| \frac{z_1 - z_2}{1 - \bar{z}_1 z_2} \right|.$$

Proof. Write $b_0 = \rho e^{i\beta}$, $\rho > 1$, and consider a single-valued analytic branch of

$$\phi(z) = \log\left(e^{-i\beta} F(z)\right) = c_0 + c_1 z + \ldots,$$

where $c_0 = \log \rho > 0$. Note $|F(z)| > 1$ means that $\mathcal{R}e\left(\phi(z)\right) > 0$. Hence for $\phi(z) = u + iv$ and $\psi(z) = \frac{\phi(z) - c_0}{\phi(z) + c_0}$, we have

$$|\psi(z)|^2 = \left| \frac{(u - c_0)^2 + v^2}{(u + c_0)^2 + v^2} \right| < 1, \qquad z \in U.$$

Since $\psi(0) = 0$, the Schwarz Lemma is applicable, giving

$$|\psi'(0)| = \left| \frac{c_1}{2c_0} \right| \le 1.$$

As a consequence,

$$|b_1| \le 2|b_0| \log |b_0|,$$

as desired.

Moreover, if $|z| = t$, then $|\psi(z)| \le t$ and so

$$|\phi(z)| - |c_0| \le |\phi(z) - c_0| \le t|\phi(z) + c_0| \le t(|\phi(z)| + c_0).$$

Therefore

$$|\phi(z)| \le c_0 \frac{1+t}{1-t},$$

implying

$$|F(z)| \le e^{|\phi(z)|} \le e^{\frac{1+t}{1-t} \log |F(0)|} = |F(0)|^{(1+t)/(1-t)}.$$

This inequality can then be applied to the function

$$G(z) = F\left(\frac{z + z_1}{1 + \bar{z}_1 z}\right),$$

since $|G(z)| > 1$, $z \in U$. Setting

$$z_2 = \frac{z + z_1}{1 + \bar{z}_1 z}, \quad \text{i.e.,} \quad z = \frac{z_2 - z_1}{1 - \bar{z}_1 z_2},$$

we deduce that

$$|F(z_2)| = |G(z)| \le |G(0)|^{(1+t)/(1-t)} = |F(z_1)|^{(1+t)/(1-t)}$$

for $t = |z|$, proving the lemma.

We are now ready to prove Theorem 3.6.10. In fact, if A and r_0 are the constants of Lemma 3.6.11, let us first suppose that $|a_0| \le e^A$. Then by Theorem 3.6.2

$$|a_1| \le M(1 + |a_0|^2) \le 2Me^A \cdot \max(1, |a_0|),$$

and the desired bound is achieved with $C = Me^A$.

On the other hand, if $|a_0| > e^A$, define $\rho > 0$ to be the largest radius for which

$$|f(z)| > 1,$$

in $|z| < \rho$. If $\rho = 1$, then by Lemma 3.6.12

$$|a_1| \le 2|a_0| \log |a_0|,$$

again yielding the desired bound with $C = 0$. In the event $\rho < 1$, let $r > 0$ be the largest radius for which $|f(z)| > e^A$ in $|z| < r$. Hence $|f(re^{i\theta})| = e^A$ for some θ.

Next observe that Lemma 3.6.12 can be applied to the function

$$F(z) = f(\rho z), \qquad |z| < 1,$$

so that for $z_1 = \frac{r}{\rho} e^{i\theta}$, $z_2 = 0$, $t = \frac{r}{\rho}$, we obtain

$$e^A = |F(z_1)| \ge |a_0|^{(\rho - r)/(\rho + r)}$$

and

$$\frac{\rho - r}{\rho + r} \log |a_0| \leq A.$$

Now, in view of the definition of ρ, there is some θ_1 for which $|f(\rho e^{i\theta_1})| = 1$; in view of the definition of r,

$$|f(re^{i\theta_1})| \geq e^A.$$

Then from Lemma 3.6.11

$$\frac{\rho - r}{1 - \rho r} \geq r_0, \qquad r_0 > 0,$$

and from the foregoing analysis

$$A \geq \frac{\rho - r}{\rho + r} \log |a_0| \geq \frac{r_0}{2}(1 - \rho r) \log |a_0|,$$

implying

$$1 - \rho \leq 1 - \rho r \leq \frac{2A}{r_0 \log |a_0|}.$$

From Lemma 3.6.12 we also obtain

$$|F'(0)| = \rho |f'(0)| \leq 2|a_0| \log |a_0|,$$

which in turn yields

$$
\begin{aligned}
|a_1| &\leq \frac{2}{\rho}|a_0| \log |a_0| = 2|a_0| \log |a_0| + \frac{2(1-\rho)}{\rho}|a_0| \log |a_0| \\
&\leq 2|a_0| \log |a_0| + \frac{4A}{r_0 \rho}|a_0|.
\end{aligned}
$$

Since $\rho \geq r_0$, we conclude that

$$|a_1| \leq 2\mu(\log \mu + C),$$

as required.

Finally, we must establish the bound on $M(r, f)$. Toward this end consider the translate of $f \in \mathcal{F}$,

$$g(z) = f\left(\frac{z + z_0}{1 + \bar{z}_0 z}\right) = f(z_0) + (1 - |z_0|^2)f'(z_0)z + \dots, \qquad z_0 \in U.$$

If $|f(z_0)| \geq 1$, then by the preceding bound on $|g'(0)|$ (as $g \in \mathcal{F}$),

$$(1 - |z_0|^2)|f'(z_0)| \leq 2|f(z_0)|\left(\log |f(z_0)| + C\right).$$

Assuming

$$M(r, f) = |f(re^{i\theta}| \geq 1$$

(for otherwise, taking $C > 0$, we are done), define r_1 to be the lower bound of all positive radii for which

$$|f(te^{i\theta})| \geq 1, \qquad r_1 \leq t \leq r.$$

Certainly either $r_1 = 0$ or $r_1 > 0$ and $|f(r_1 e^{i\theta})| = 1$, but in both cases

$$|f(r_1 e^{i\theta})| \leq \mu = \max(1, |a_0|).$$

Furthermore, for $r_1 \leq t \leq r$,

$$|f'(te^{i\theta})| \leq \frac{2}{1-t^2} |f(te^{i\theta})| \left(\log |f(te^{i\theta})| + C \right),$$

implying

$$\frac{\partial}{\partial t} \log \left(\log |f(te^{i\theta})| + C \right) \leq \frac{2}{1-t^2}.$$

If we integrate both sides from $t = r_1$ to $t = r$, the result becomes

$$\frac{\log \left(M(r,f) + C \right)}{\log |f(r_1 e^{i\theta})| + C} \leq \left(\frac{1-r_1}{1+r_1} \right) \frac{1+r}{1-r},$$

and therefore

$$\log M(r,f) + C \leq (\log \mu + C) \frac{1+r}{1-r},$$

that is,

$$\log M(r,f) \leq \frac{1+r}{1-r} \log \mu + \frac{2Cr}{1-r},$$

and the proof is now complete.

The fact that the bounds in the above formulation are sharp can be readily shown. In fact, fix an arbitrary constant $C \in \mathbf{R}$ and define

$$\mathcal{F}_C = \{ f_\alpha(z) = e^{-C + \alpha(1+z)/(1-z)} : \alpha > 0, \ |z| < 1 \}.$$

Since $\mathcal{R}e \left(\alpha(1+z)/(1-z) \right) > 0$, $|f_\alpha(z)| > e^{-C}$, so that $\mathcal{F}_C$ is normal and clearly invariant. Moreover, for any $\alpha > C$,

$$a_0 = e^{-C+\alpha} = \mu,$$
$$a_1 = 2\alpha e^{-C+\alpha} = 2\mu(\log \mu + C)$$

and

$$M(r,f) = e^{-C+\alpha(1+r)/(1-r)} = e^{(-C+\alpha)(1+r)/(1-r) + 2Cr/(1-r)}$$
$$= \mu^{(1+r)/(1-r)} e^{2Cr/(1-r)},$$

as required.

Observe that in the classical Schottky Theorem, the family of analytic functions in U which omit 0 and 1 is normal (by the FNT) and invariant, so that the above result is applicable. Moreover, if $|a_0| \geq 1$, then

$$|a_1| \leq 2|a_0|(|\log|a_0|| + C), \tag{3.6}$$

and for $|a_0| < 1$, the same inequality is seen to hold by considering $\frac{1}{f(z)}$. An application of (3.6) to $f(Rz)$, where

$$f(z) = a_0 + a_1 z + \ldots \qquad (a_1 \neq 0)$$

is analytic in $|z| < R$, $f(z) \neq 0, 1$, gives

$$|a_1|R \leq 2|a_0|(|\log|a_0|| + C).$$

Thus, we obtain for a bound in Landau's theorem

$$M(a_0, a_1) = \frac{2|a_0|(|\log|a_0|| + C)}{|a_1|}.$$

For explicit best bounds in Schottky's theorem, see Hempel [1979, 1980]. Also, Hempel [1979] and Lai [1979] have given a sharp value for the afore-mentioned C in Landau's theorem: $C = \Gamma\left(\frac{1}{4}\right)^4/(4\pi^2) = 4.3768796\ldots$.

3.7 Asymptotic Values

We discuss here but a small part of this important subject. For further details refer to Segal [1981], Chapter 5.

Definition 3.7.1 *Let $f(z)$ be meromorphic and $w \in \widehat{\mathbb{C}}$. If $f(z) \to w$ as $|z| \to \infty$ along some continuous path γ, then w is an* **asymptotic value** *of $f(z)$ and γ is called a* **path of determination**.

For example, $f(z) = e^z$ has two asymptotic values 0 and ∞ and the negative/positive real axis is a path of determination of $0/\infty$, respectively. It is well-known that a *Picard exceptional value* (that is, a value taken only finitely often) is an asymptotic value (cf. Segal [1981], p. 232).

The following is another version of Julia's theorem [1924], which exploits the notion of normal families, in this instance of meromorphic functions.

Theorem 3.7.2 *Let $f(z)$ be a nonrational meromorphic function in $\mathbb{C}$ having the asymptotic value w. Let $\gamma = \{\sigma(t) : 0 \leq t < \infty, \ \sigma(0) > 0\}$ be a Jordan path extending to ∞. Then for some point $z_0 \in \mathbb{C}$ and any arbitrarily small $r > 0$, $f(z)$ attains every value in $\mathbb{C}$, with at most two exceptions, infinitely often in the band*

$$\bigcup_{t \geq 0} D\big(z_0\sigma(t); r|\sigma(t)|\big).$$

Proof. Define
$$\mathcal{F} = \{f_t(z) = f(z\sigma(t)), 0 \le t < \infty\},$$
and let $\mathcal{A}$ be the annulus: $\{0 < R_1 < |z| < R_2 < \infty\}$ with $\sigma(0) \in \mathcal{A}$. Suppose that $\mathcal{F}$ is normal in $\mathcal{A}$. Then there is a subsequence $\{f_{t_n}\} \subseteq \mathcal{F}$ ($t_1 < t_2 < \ldots < t_n < \ldots$, $t_n \to \infty$) converging spherically uniformly on compact subsets of $\mathcal{A}$ to a meromorphic function $\phi(z)$ or to ∞.

If the limit is $\phi(z)$, let Γ be a path of determination from 0 to ∞ with $f(z) \to w$ as $|z| \to \infty$ along Γ. Choose a circle $C_R : |z| = R$, $R_1 < R < R_2$. Then for $n = 1, 2, 3 \ldots$, there is some point $\zeta_n \in C_R$ such that $\zeta_n \sigma(t_n) \in \Gamma$. Consequently, for $\varepsilon > 0$

$$\chi\big(f_{t_n}(\zeta_n), w\big) = \chi\big(f(\zeta_n\sigma(t_n)), w\big) < \frac{\varepsilon}{3},$$

for sufficiently large n. If $\zeta \in C_R$ is a limit point of $\{\zeta_n\}$, since $\mathcal{F}$ is spherically equicontinuous on C_R, there exists $\delta > 0$ such that

$$\chi\big(f_{t_n}(\zeta), f_{t_n}(\zeta_n)\big) < \frac{\varepsilon}{3},$$

whenever $|\zeta - \zeta_n| < \delta$. From the convergence $f_{t_n} \to \phi$, we have

$$\chi\big(\phi(\zeta), w\big) \le \chi\big(\phi(\zeta), f_{t_n}(\zeta)\big) + \chi\big(f_{t_n}(\zeta), f_{t_n}(\zeta_n)\big) + \chi\big(f_{t_n}(\zeta_n), w\big) < \varepsilon,$$

for sufficiently large n, i.e., $\phi(\zeta) = w$. Since R is arbitrary, $R_1 < R < R_2$, there is a point $\zeta = \zeta_R$ on each C_R satisfying $\phi(\zeta_R) = w$. Hence $\phi(\zeta) \equiv w$ in $\mathcal{A}$, and

$$\lim_{n\to\infty} f\big(z\sigma(t_n)\big) = \lim_{n\to\infty} f_{t_n}(z) = \phi(z) = w, \qquad z \in \mathcal{A}.$$

However, ∞ is an isolated essential singularity of $f(z)$, so that invoking the Casorati-Weierstrass Theorem (cf. Palka [1991], p. 321) we have a contradiction.

If $\{f_{t_n}\}$ converges to ∞ in $\mathcal{A}$, we similarly obtain a contradiction.

It follows that $\mathcal{F}$ is not normal in $\mathcal{A}$, and there exists a point $z_0 \in \mathcal{A}$ such that $\mathcal{F}$ is not normal in any disk $D(z_0; r) \subseteq \mathcal{A}$. Define

$$\Delta_r = \bigcup_{t \ge 0}\{\zeta : \zeta = z\sigma(t), z \in D(z_0; r)\}.$$

For $\zeta \in \Delta_r$, $f(\zeta) = f(z\sigma(t)) = f_t(z)$, $z \in D(z_0; r)$. Since $\{f_t\}$ is not normal in any $D(z_0; r)$, $f(\zeta)$ must take every value in $\mathbb{C}$, for $\zeta \in \Delta_r$, with at most two exceptions by the FNT (§3.2). Clearly the values taken are done so infinitely often.

Corollary 3.7.3 *Under the assumptions of the theorem, there is at least one ray α through the origin such that in every angular domain bisected by α, $f(z)$ takes every value in $\widehat{\mathbb{C}}$, with at most two exceptions, infinitely often.*

Indeed, we can take γ to be a ray whose extension passes through the origin. The ray α is again called a *direction* (*line*) *of Julia* (cf. §2.8).

Corollary 3.7.4 *If under the conditions of the theorem,* $f(z)$ *is entire, then the conclusion holds for values in* $\mathbb{C}$ *with at most one omitted value.*

In this instance, ∞ is already an omitted value.

Corollary 3.7.5 *If* $f(z)$ *is entire, there is at least one ray* α *through the origin such that in very angular domain bisected by* α, $f(z)$ *takes every value in* $\mathbb{C}$, *with at most one exception, infinitely often.*

Example 3.7.6 From the periodic nature of the function $f(z) = e^z$, we readily find that $\{\arg z = \pm\pi/2\}$ are its lines of Julia. Moreover, these are the only two lines of Julia.

3.8   Linear Fractional Transformations

The subject of linear fractional (bilinear or Möbius) transformations is discussed in depth from the standpoint of discontinuous groups in §5.5. Here we content ourselves with examining some normality criteria for these transformations. Further criteria can be found in §5.5.

Consider the family $\mathcal{L}_M$ of linear fractional transformations given by

$$w = T(z) = \frac{az + b}{cz + d}, \qquad ad - bc = 1, \quad |a|, |b|, |c|, |d| < M < \infty,$$

where $a, b, c, d \in \mathbb{C}$, $z \in \widehat{\mathbb{C}}$. We claim that: $\mathcal{L}_M$ *is normal in the extended complex plane.* To this end, note that from any $\{T_n\} \subseteq \mathcal{L}_M$, a subsequence $\{T_\nu\}$ may be extracted, for which

$$a_\nu \to a_0, \quad b_\nu \to b_0, \quad c_\nu \to c_0, \quad d_\nu \to d_0, \qquad a_0 d_0 - b_0 c_0 = 1.$$

If K is a closed disk in $\mathbb{C}$ with $-\frac{d_0}{c_0} \notin K$, then

$$\frac{a_\nu z + b_\nu}{c_\nu z + d_\nu} \to \frac{a_0 z + b_0}{c_0 z + d_0},$$

uniformly on K. On the other hand, noting that $\frac{d_0}{c_0} \neq \frac{b_0}{a_0}$, choose a closed disk K containing the point $-\frac{d_0}{c_0}$ but excluding the point $-\frac{b_0}{a_0}$. Then

$$\frac{c_\nu z + d_\nu}{a_\nu z + b_\nu} \to \frac{c_0 z + d_0}{a_0 z + b_0}$$

uniformly on K. Finally, working in a neighbourhood of ∞, we write

$$T_\nu(z) = \frac{a_\nu + b_\nu \frac{1}{z}}{c_\nu + d_\nu \frac{1}{z}}$$

and proceed similarly as above, thus establishing the normality of $\mathcal{L}_M$ by Theorem 3.1.3.

Next, let us extend the preceding analysis somewhat. For an arbitrary $T \in \mathcal{L}_M$, denote

$$w_0 = \frac{b}{d}, \quad w_1 = \frac{a+b}{c+d}, \quad w_\infty = \frac{a}{c}$$

as the images of the points $z = 0, 1, \infty$, respectively. Then the product of the three chordal distances of these points from one another yields the expression

$$\chi(w_0, w_1) \cdot \chi(w_1, w_\infty) \cdot \chi(w_\infty, w_0)$$
$$= \frac{|ad - bc|^3}{(|a|^2 + |c|^2)(|b|^2 + |d|^2)(|a + b|^2 + |c + d|^2)}.$$

Since $T \in \mathcal{L}_M$, the constraints on the coefficients a, b, c, d imply that

$$\chi(w_0, w_1) \cdot \chi(w_1, w_\infty) \cdot \chi(w_\infty, w_0) > \frac{1}{N^2},$$

for some positive number N.

Suppose that we now consider the converse situation of a family $\mathcal{L}$ of linear fractional transformations

$$w = T(z) = \frac{az + b}{cz + d}, \qquad ad - bc = 1,$$

for which the three chordal distances as given above satisfy

$$\chi(w_0, w_1) \cdot \chi(w_1, w_\infty) \cdot \chi(w_\infty, w_0) > \frac{1}{N^2}.$$

We infer that

$$(|a|^2 + |c|^2)(|b|^2 + |d|^2)(|a + b|^2 + |c + d|^2) < N^2,$$

which yields the inequalities

$$|a| \cdot |d| \cdot |a + b| < N, \qquad |a| \cdot |b| \cdot |c + d| < N.$$

Coupled with the identity

$$a = a[d(a + b) - b(c + d)]$$
$$= ad(a + b) - ab(c + d),$$

these inequalities give the result that $|a| < 2N$. In similar fashion, it is found that $|b|, |c|, |d| < 2N$. This then means: $\mathcal{L}$ is a normal family, as was determined previously. Carathéodory [1958, Vol. I], p. 193, has remarked that N^2 must be taken not less than $8/\sqrt{27}$ in order to ensure that $\mathcal{L} \neq \emptyset$. See §4.1 for a related result.

From the foregoing considerations, we can deduce

Theorem 3.8.1 *Let $\mathcal{F}$ be a family of meromorphic functions in a domain Ω. Suppose that to each $f \in \mathcal{F}$ there is associated four complex numbers a, b, c, d satisfying $ad - bc = 1$, $|a|, |b|, |c|, |d| < M$. Then the corresponding family $\mathcal{G}$ of functions*

$$g(z) = \frac{af(z) + b}{cf(z) + d},$$

is normal in Ω whenever $\mathcal{F}$ is, and vice versa.

Before we present an important application (Theorem 3.9.1) of this theorem, it is necessary to have the following meromorphic version of the

Hurwitz Theorem *Let $\{f_n\}$ be a sequence of meromorphic functions on a domain Ω which converge spherically uniformly on compact subsets to a function f (which may be $\equiv \infty$). If each f_n is zero-free in Ω, then f is either zero-free in Ω or $f \equiv 0$.*

Proof. Suppose that $f \not\equiv 0$ and consider any point $z_0 \in \Omega$. If $f(z_0) \neq \infty$, then by Theorem 3.1.3, there is a closed disk about z_0 in which f is analytic and in which $f_n \to f$ uniformly. Therefore f is zero-free in a neighbourhood of z_0 by an appeal to the Hurwitz Theorem for analytic functions (§1.4).

Corollary 3.8.2 *Let $\mathcal{F}$ be a normal family of meromorphic functions in Ω such that $f(z) \neq a$ for all $f \in \mathcal{F}$ and some $a \in \mathbb{C}$. Then no limit function F of a normally convergent sequence in $\mathcal{F}$ can attain the value a unless $F \equiv a$.*

3.9   Univalent Functions

As is the case with analytic functions, meromorphic functions that are one-to-one are also called *univalent* or *schlicht*. The linear fractional transformations are univalent functions on $\widehat{\mathbb{C}}$. Recall (Corollary 1.4.1) that if a sequence of univalent analytic functions converges uniformly on compact subsets of a domain to a nonconstant analytic function f, then f is necessarily univalent. The analogous result in the case of univalent meromorphic functions is

Theorem 3.9.1 *Suppose that $\{f_n\}$ is a sequence of meromorphic functions on a domain Ω which converge spherically uniformly on compact subsets to a function f. If each f_n is univalent, then either f is univalent or f is identically constant.*

Proof. Assume $f \not\equiv \infty$ (therefore, f is meromorphic in Ω) and fix a point $z_0 \in \Omega$. To each function f_n associate a function g_n given by

$$g_n(z) = f_n(z) - f_n(z_0) \qquad \text{if } |f_n(z_0)| \le 1,$$

$$g_n(z) = \frac{\frac{1}{f_n(z_0)} f_n(z) - 1}{\left(1 - \frac{1}{f_n(z_0)}\right) f_n(z) + 1} \qquad \text{if } 1 < |f_n(z_0)| < \infty,$$

$$g_n(z) = -\frac{1}{f_n(z)} \qquad \text{if } f_n(z_0) = \infty.$$

Then $g_n(z_0) = 0$, whereas $g_n(z) \ne 0$ for all $z \in \Omega_0 = \Omega - \{z_0\}$. Let us write each g_n in the more general form

$$g_n(z) = \frac{a_n f_n(z) + b_n}{c_n f_n(z) + d_n},$$

and note that the coefficients from the above relations are all bounded in modulus by 2 with $a_n d_n - b_n c_n = 1$. In view of Theorem 3.8.1 there is a subsequence $\{f_{n_k}\} \subseteq \{f_n\}$ such that $a_{n_k} \to a$, $b_{n_k} \to b$, $c_{n_k} \to c$, $d_{n_k} \to d$, and the corresponding functions $\{g_{n_k}\}$ converge spherically uniformly on compact subsets to

$$g(z) = \frac{a f(z) + b}{c f(z) + d}.$$

Now, $g(z) \ne 0$ for all $z \in \Omega_0$ by the preceding Hurwitz Theorem, unless of course $g(z) \equiv 0$. This latter possibility obtains only if $f \equiv$ constant in Ω. We conclude that if $f \not\equiv$ constant, then $f(z) \ne f(z_0)$ for all $z \in \Omega_0$, and the proof is complete.

4

Bloch Principle

The origin of the Bloch Principle can seemingly be traced back to Bloch's dictum, *"Nihil est in infinito quod non prius fuerit in finito"* found on p. 2 of his 1926 monograph, as well as on p. 84 of [1926b]. It may be translated as: *Nothing exists in the infinite plane that has not been previously done in the finite disk.* In modern parlance it is the hypothesis that a family of analytic (meromorphic) functions which have a common property $\mathcal{P}$ in a domain Ω will in general be a normal family if $\mathcal{P}$ reduces an analytic (meromorphic) function in $\mathbb{C}$ to a constant. The property of omitting two (resp. three) given values of the FNT is one such example. However, the connection between the modern Bloch Principle and Bloch's original utterance is tenuous at best.

A formalization of the Bloch Principle, due to Robinson and Zalcman, is presented, from which we produce almost trivial proofs of some standard results such as the FNT. Unfortunately, there are exceptions to the general Bloch Principle and we give two counterexamples due to Rubel. Moreover, the converse of the Bloch Principle is valid and is proved under rather weak assumptions. As it turns out, there are a great many instances in which the Bloch Principle holds, and it has become a valuable heuristic tool. In this regard, Drasin has produced a wide-ranging body of results embracing the Bloch Principle using the Nevanlinna theory. This approach dates back to the work of Cartan [1928], Bureau [1932], and Miranda [1935], and modern practitioners such as Drasin, Yang, Schwick, among others, have embraced somewhat different points of view.

4.1 Robinson-Zalcman Heuristic Principle

The Bloch Principle was noted in an address by Abraham Robinson [1973] as one of twelve mathematical problems requiring further consideration. He gave a formulation which has been made rigorous by Zalcman [1975], exploiting the work of Lohwater and Pommerenke [1973].

What Zalcman derived is the following important characterization of normality, based on Theorem 1 of the aforementioned work.

Zalcman Lemma *Let $\mathcal{F}$ be a family of analytic (meromorphic) functions in U. Then $\mathcal{F}$ is not normal in U if and only if there exist (i) a number r*

with $0 < r < 1$; (ii) points z_n satisfying $|z_n| < r$; (iii) functions $f_n \in \mathcal{F}$; (iv) positive numbers $\rho_n \to 0$ as $n \to \infty$; such that

$$f_n(z_n + \rho_n \zeta) \to g(\zeta) \qquad \text{as } n \to \infty, \tag{4.1}$$

uniformly (spherically uniformly) on compact subsets of $\mathbb{C}$, where g is a nonconstant entire (meromorphic) function in $\mathbb{C}$.

Proof. Assume $\mathcal{F}$ is not normal in U. By Marty's theorem (§3.3) there exists a number r_0, $0 < r_0 < 1$, a sequence of functions $\{f_n\} \subseteq \mathcal{F}$, and a sequence of points z_n' in $\{|z| \le r_0\}$ such that $f_n^\#(z_n') \to \infty$. For fixed r with $r_0 < r < 1$, since $f^\#(z)$ is continuous on $|z| \le r$, we define

$$M_n = \max_{|z| \le r} \left(1 - \frac{|z|^2}{r^2}\right) f_n^\#(z) = \left(1 - \frac{|z_n|^2}{r^2}\right) f_n^\#(z_n). \tag{4.2}$$

It is evident that $M_n \to \infty$. Next, set

$$\rho_n = \frac{1}{M_n}\left(1 - \frac{|z_n|^2}{r^2}\right) = \frac{1}{f_n^\#(z_n)} \to 0 \qquad \text{as } n \to \infty, \tag{4.3}$$

so that

$$\frac{\rho_n}{r - |z_n|} = \frac{r + |z_n|}{r^2 M_n} \le \frac{2}{r M_n} \to 0 \qquad \text{as n} \to \infty. \tag{4.4}$$

Then the functions

$$g_n(\zeta) = f_n(z_n + \rho_n \zeta)$$

are defined on $|\zeta| < R_n = \frac{r - |z_n|}{\rho_n}$, and $R_n \to \infty$ as $n \to \infty$. Hence by (4.3)

$$g_n^\#(0) = \rho_n f_n^\#(z_n) = 1$$

for each n. Moreover, taking a fixed R with $|\zeta| \le R < R_n$, then we have $|z_n + \rho_n \zeta| < r$, and together with (4.2) and (4.3) obtain

$$
\begin{aligned}
g_n^\#(\zeta) &= \rho_n f_n^\#(z_n + \rho_n \zeta) \\
&\le \frac{\rho_n M_n}{1 - \frac{|z_n + \rho_n \zeta|^2}{r^2}} \le \frac{r + |z_n|}{r + |z_n| + \rho_n R} \cdot \frac{r - |z_n|}{r - |z_n| - \rho_n R} \\
&\le \frac{r - |z_n|}{r - |z_n| - \rho_n R} \to 1 \qquad \text{as } n \to \infty,
\end{aligned}
$$

by (4.4). It follows that $\{g_n\}$ is normal in $|\zeta| < R$ by Marty's theorem, so that a subsequence $\{g_{n_k}\}$ converges spherically uniformly to a function g on compact subsets of $\mathbb{C}$. Since

$$g^\#(0) = \lim_{k \to \infty} g_{n_k}^\#(0) = 1,$$

g is a nonconstant entire (meromorphic) function, as desired.

Conversely, suppose that conditions (i)–(iv) hold as well as (4.1) but $\mathcal{F}$ is normal. Then there exists a constant $M > 0$ such that

$$\max_{|z| \leq \frac{1+r}{2}} f^{\#}(z) \leq M, \qquad f \in \mathcal{F},$$

by Marty's theorem. For a fixed $\zeta \in \mathbb{C}$, $|z_n + \rho_n\zeta| \leq \frac{1+r}{2}$ for sufficiently large n, implying $\rho_n f_n^{\#}(z_n + \rho_n\zeta) \leq \rho_n M$. Therefore, by (4.1),

$$g^{\#}(\zeta) = \lim_{k \to \infty} \rho_n f_n^{\#}(z_n + \rho_n\zeta) = 0.$$

Since ζ is arbitrary, $g \equiv$ constant (possibly infinity), a contradiction, which completes the proof.

For generalizations of this key result, cf. Pang [1989], Chen and Gu [preprint].

In the sequel we shall employ Robinson's notation denoting a function together with its domain, such as $\langle f, \Omega \rangle$; if $\Omega \neq \Omega'$ we distinguish between $\langle f, \Omega \rangle$ and $\langle f, \Omega' \rangle$. The following characterization is essentially due to Robinson [1973], and refined by Zalcman [1975]; cf. also Zalcman [1982].

Definition 4.1.1 *A property $\mathcal{P}$ of analytic (meromorphic) functions is called* **normal** *whenever $\mathcal{P}$ satisfies the following conditions ($\langle f, \Omega \rangle \in \mathcal{P}$ means f has property $\mathcal{P}$ on Ω):*

(i) *If $\langle f, \Omega \rangle \in \mathcal{P}$ and $\Omega' \subseteq \Omega$, then $\langle f, \Omega' \rangle \in \mathcal{P}$.*

(ii) *If $\langle f, \Omega \rangle \in \mathcal{P}$ and $\phi(z) = az + b$, then $\langle f \circ \phi, \phi^{-1}(\Omega) \rangle \in \mathcal{P}$.*

(iii) *Let $\langle f_n, \Omega_n \rangle \in \mathcal{P}$, where $\Omega_1 \subseteq \Omega_2 \subseteq \ldots$ and $\mathbb{C} = \cup_{n=1}^{\infty} \Omega_n$. If $f_n \to f$ uniformly (spherically uniformly) on compact subsets of $\mathbb{C}$, then $\langle f, \mathbb{C} \rangle \in \mathcal{P}$.*

(iv) *If $\langle f, \mathbb{C} \rangle \in \mathcal{P}$, then $f \equiv$ constant.*

We remark that (ii) need only be valid for $0 < a \leq 1$. A version of the Bloch Principle can then be formulated as the

Robinson-Zalcman Heuristic Principle *If $\mathcal{P}$ is a normal property, then, for any domain $\Omega \subseteq \mathbb{C}$, the family*

$$\mathcal{F} = \{f : \langle f, \Omega \rangle \in \mathcal{P}\}$$

is normal in Ω.

Proof. Let us suppose not. Then $\mathcal{F}$ is not normal in some disk $\Delta \subseteq \Omega$. For any $f \in \mathcal{F}$, $\langle f, \Delta \rangle \in \mathcal{P}$ by (i). By (ii) we may take $\Delta = U : |z| < 1$. Next apply the Zalcman Lemma to $\mathcal{F}$ (and U), with r, $|z_n|$, ρ_n, as given therein.

Letting $R_n = \frac{r - |z_n|}{\rho_n}$, $R_n \to \infty$, and we may assume $\{R_n\}$ is an increasing sequence by considering a subsequence if necessary. For the functions

$$g_n(\zeta) = f_n(z_n + \rho_n \zeta)$$

defined on $\Omega_n : |\zeta| < R_n$, we have $|z_n + \rho_n \zeta| < r < 1$, so that $\langle g_n, \Omega_n \rangle \in \mathcal{P}$ for each n by (ii). Since $\cup_{n=1}^{\infty} \Omega_n = \mathbb{C}$ and $g_n \to g$ normally in $\mathbb{C}$, $\langle g, \mathbb{C} \rangle \in \mathcal{P}$ by (iii). A fortiori $g \equiv$ constant by (iv), and this contradicts the assertion of the Zalcman Lemma. We conclude that $\mathcal{F}$ is normal in Ω.

For a nonstandard analysis proof, see Stroyan and Luxemburg [1976], pp. 219–220. Furthermore, the notion of a normal property has been extended by Pang, replacing (ii) with

(ii′) If k is a real number $(-1 < k < 1)$, $\langle f, \Omega \rangle \in \mathcal{P}$ and $\phi(z) = az + b$, then $\langle \frac{f \circ \phi}{a^k}, \phi^{-1}(\Omega) \rangle \in \mathcal{P}$.

Exploiting (ii′), Pang [1989, 1990] deduced the Heuristic Principle along similar lines to that above and showed the domain $(-1, 1)$ for k to be the largest possible. This extension has a number of important consequences insofar as results involving derivatives may be obtained without recourse to Nevanlinna theory (cf. Theorems 4.4.18, 4.4.19, and 4.5.2).

A stronger form of the Heuristic Principle has been obtained by Ros [preprint] in his study of the Gauss map of minimal surfaces.

Example 4.1.2 Define $\langle f, \Omega \rangle \in \mathcal{P}_M$ if $f(z)$ is analytic in Ω and

$$D_{\Omega}(f) = \int_{\Omega} |f'(z)|^2 \, dx \, dy \leq M < \infty.$$

Then (i)–(iv) of Definition 4.1.1 are readily verified, with (iii) a consequence of Fatou's Lemma, and (iv) is worked out in the discussion following Example 5.4.9. Thus $\mathcal{F}_M = \{f : \langle f, \Omega \rangle \in \mathcal{P}_M\}$ is normal in Ω, as was previously shown in Example 2.2.7 by other means.

As a further illustration we give an elementary proof of the Fundamental Normality Test (§2.7). Indeed, define the property $\mathcal{P}$: f *is analytic and either constant or omits the values a and b*. Then conditions (i) and (ii) are clearly satisfied, and (iii) follows from the Hurwitz Theorem (§1.4). Since (iv) is just Picard's first theorem, we conclude that $\mathcal{P}$ is a normal property and $\mathcal{F} = \{f : \langle f, \Omega \rangle \in \mathcal{P}\}$ is a normal family.

There is another splendid version of the FNT for meromorphic functions (§3.2) due to Carathéodory [1960, Vol. II], p. 202.

Extended FNT *Let $\mathcal{F}$ be a family of meromorphic functions on a domain Ω. Suppose that each $f \in \mathcal{F}$ omits three distinct values a, b, c (which may depend on f) satisfying $\chi(a, b) \cdot \chi(b, c) \cdot \chi(c, a) \geq \varepsilon$, for some $\varepsilon > 0$ (which is independent of f). Then $\mathcal{F}$ is normal in Ω.*

Proof. Define $\mathcal{P}$: f omits three values a, b, c satisfying

$$\chi(a,b) \cdot \chi(b,c) \cdot \chi(c,a) \geq \varepsilon > 0.$$

Then (i) and (ii) are obviously satisfied. For (iii), suppose $\chi(f_n, f) \to 0$ uniformly on compact subsets of $\mathbb{C}$ and that f_n omits points a_n, b_n, c_n, where $\chi(a_n, b_n) \cdot \chi(b_n, c_n) \cdot \chi(c_n, a_n) \geq \varepsilon$. As a constant function trivially satisfies $\mathcal{P}$, we may assume f is nonconstant. Moreover, by the compactness of the Riemann sphere, there are limit points a, b, c, a subsequence, again denoted by $\{f_n\}$, and corresponding $\{a_n\}, \{b_n\}, \{c_n\}$, with $\chi(a_n, a) \to 0$, $\chi(b_n, b) \to 0$, $\chi(c_n, c) \to 0$. Thus $\chi(a, b) \cdot \chi(b, c) \cdot \chi(c, a) \geq \varepsilon$ by continuity. It remains to show that f omits the values a, b, c. On the contrary, suppose that $f(z_0) = a$. If $a \neq \infty$, then in some closed disk $K(z_0; r)$, $f(z)$ is analytic and bounded, with $f_n - a_n \to f - a$ uniformly on $K(z_0; r)$. By the Hurwitz Theorem, since f is nonconstant, $f_n(z) - a_n$ has a zero in $K(z_0; r)$ for all n sufficiently large, a contradiction. When $a = \infty$, $\frac{1}{f_n} \to \frac{1}{f}$ uniformly in $K(z_0; r)$, yielding a contradiction as before. Property (iv) again follows from Picard's first theorem, completing the proof.

The preceding generalizes an earlier result of Montel [1927], p. 73: *If $\mathcal{F}$ is a family of analytic functions such that each $f \in \mathcal{F}$ omits two distinct values a, b (which may depend on f) satisfying $|a| < M$, $|b| < M$, $|a - b| \geq \varepsilon > 0$ (with ε independent of f), then $\mathcal{F}$ is normal.* Montel considers the corresponding functions $g(z) = (f(z) - a)/(b - a)$ which omit 0 and 1. The obvious details are left to the reader.

Another generalization of the FNT is Theorem 3.2.2, which itself admits the following significant improvement.

Theorem 4.1.3 *Let $\mathcal{F}$ be a family of meromorphic functions in a domain Ω. If for each $f \in \mathcal{F}$, the poles are of multiplicity $\geq h$, the zeros are of multiplicity $\geq k$, and the zeros of $f(z) - 1$ are of multiplicity $\geq \ell$, with*

$$\frac{1}{h} + \frac{1}{k} + \frac{1}{\ell} < 1,$$

then $\mathcal{F}$ is normal in Ω.

Indeed, it is routine to show that the conditions on $f \in \mathcal{F}$ constitute a normal property, where condition (iii) is a consequence of the Hurwitz Theorem (first invoking Theorem 3.1.3 in the case of a pole), and (iv) is just Corollary 1.8.8. The result then follows by the Heuristic Principle. (For a proof from the point of view of Nevanlinna theory, see Theorem 4.4.5).

In an analogous fashion one can prove the more general (cf. Bloch [1926a], p. 49, Valiron [1929], p. 35, Yang and Chang [1966]).

Theorem 4.1.4 *Let $\mathcal{F}$ be a family of meromorphic functions on a domain Ω such that each zero of $f(z) - a_\nu$, $a_\nu \in \widehat{\mathbb{C}}$ has multiplicity $\geq m_\nu$ (≥ 2),*

$\nu = 1, 2, \ldots, q.$ *If*

$$\sum_{\nu=1}^{q} \left(1 - \frac{1}{m_\nu} \right) > 2,$$

then $\mathcal{F}$ is normal in Ω.

Recall from Example 1.8.7, that if $f(z)$ is meromorphic in $\mathbb{C}$ with

$$\sum_{\nu=1}^{q} \left(1 - \frac{1}{m_\nu} \right) > 2,$$

then $f(z)$ reduces to a constant.

The notion of a normal property can also be applied to the study of essential singularities (Minda [1985]).

Theorem 4.1.5 *If $\mathcal{P}$ is a normal property of analytic functions and $\langle f, D'(z_0; r) \rangle \in \mathcal{P}$, then f has a nonessential singularity at z_0.*

Proof. Consider z belonging to the deleted neighbourhood $D'(z_0; r)$ such that $|z - z_0| \le \frac{r}{2}$. Then $D(z; |z - z_0|) \subseteq D'(z_0; r)$, so that condition (i) defining a normal property implies $\langle f, D(z; |z - z_0|) \rangle \in \mathcal{P}$. Since $\phi(\zeta) = z + |z - z_0|\zeta$ is a linear mapping of U onto $D(z; |z - z_0|)$, condition (ii) gives $\langle f \circ \phi, U \rangle \in \mathcal{P}$. Moreover, by the Heuristic Principle, the family

$$\mathcal{G} = \{ g : \langle g, U \rangle \in \mathcal{P} \}$$

is normal in U, so that by Marty's theorem (§3.3)

$$g^{\#}(0) \le M < \infty, \qquad g \in \mathcal{G}.$$

As $(f \circ \phi)^{\#}(0) = |z - z_0| f^{\#}(z)$, we deduce that

$$\limsup_{z \to z_0} |z - z_0| f^{\#}(z) \le M.$$

However, a necessary condition for z_0 to be an essential singularity (Lehto and Virtanen [1957b]) is $\limsup_{z \to z_0} |z - z_0| f^{\#}(z) = \infty$, proving the theorem.

Converse of the Bloch Principle. The converse of the Bloch Principle is essentially the statement that: *If any family of meromorphic functions satisfying a (suitable) property $\mathcal{P}$ in an arbitrary domain is necessarily normal, then a function that is meromorphic in $\mathbb{C}$ and possesses $\mathcal{P}$ reduces to a constant.* This idea appears implicitly in Montel [1927], p. 133, and also with regard to harmonic functions in Montel [1935] (cf. §5.4). In order to prove the converse, only mild assumptions about $\mathcal{P}$ are required (cf. Stroyan and Luxemburg [1976], p. 219).

Theorem 4.1.6 *Let $\mathcal{P}$ be a property satisfying conditions (i) and (ii) of Definition 4.1.1. Assume that if $\mathcal{F}$ is any family of meromorphic functions defined in U with $\langle f, U \rangle \in \mathcal{P}$, for all $f \in \mathcal{F}$, then $\mathcal{F}$ is normal. Then any f_0 which is meromorphic in $\mathbb{C}$ with $\langle f_0, \mathbb{C} \rangle \in \mathcal{P}$ reduces to a constant.*

Proof. (Suggested by Zalcman). Let f_0 be given as above and let z_0 be an arbitrary point of $\mathbb{C}$. Define the family

$$\mathcal{F} = \{f_R(z) = f_0(z_0 + Rz) : R > 0, \ |z| < 1\}.$$

Then $\langle f_0, \mathbb{C} \rangle \in \mathcal{P}$ implies $\langle f_0, D(z_0; R) \rangle \in \mathcal{P}$ by (i) for any $R > 0$. Hence $\langle f_R, U \rangle \in \mathcal{P}$ in view of (ii) for all $f_R \in \mathcal{F}$. Therefore, $\mathcal{F}$ is normal in U by hypothesis, so that by Marty's theorem (§3.3), the values

$$f_R^{\#}(0) = R f_0^{\#}(z_0)$$

must remain bounded for all $R > 0$. We conclude that $f_0^{\#}(z_0) = 0$ and therefore $f_0 \equiv$ constant in $\mathbb{C}$.

4.2 Counterexamples

In spite of the Robinson-Zalcman formalization, the Bloch Principle as originally stated is not universally true. One difficulty encountered in applying the Heuristic Principle is the condition (ii) of linear invariance, particularly in dealing with derivatives, although this has been partly overcome in the extension by Pang [1989, 1990]. We present two counterexamples to the Bloch Principle formulated by Rubel [1986], who also discusses two others.

Counterexample 4.2.1 Let $\mathcal{P}$ be the property of the analytic function $f : f = g''$, *for some univalent function g.*

(I) If f is entire, having property $\mathcal{P}$, i.e., $\langle f, \mathbb{C} \rangle \in \mathcal{P}$, then $f = g''$ implies g is entire, univalent, and so g must be linear, i.e., $g(z) = az + b$, for some constants a, b. We deduce that $f \equiv 0$.

(II) On the other hand, define for $z \in U$

$$g_n(z) = n\left(z + \frac{z^2}{10} + \frac{z^3}{10} \right), \qquad n = 1, 2, 3, \dots .$$

Then $g_n'(z) = n\left(1 + \frac{2z}{10} + \frac{3z^2}{10}\right)$, and $\mathcal{R}e\, g_n' > 0$ for $z \in U$. According to the Noshiro-Warschawski Theorem (cf. Duren [1983], Theorem 2.16), $g_n(z)$ is univalent in U. Let

$$\mathcal{F} = \left\{ f_n(z) = g_n''(z) = n\left(\frac{1}{5} + \frac{3z}{5}\right) : z \in U, \ n = 1, 2, 3, \dots \right\}.$$

Then $f_n\left(-\frac{1}{3}\right) = 0$, whereas $f_n(z) \to \infty$ for $z \neq -\frac{1}{3}$, $z \in U$. Hence $\mathcal{F}$ is not a normal family in U and the Bloch Principle is violated in this case.

Counterexample 4.2.2 Let $\mathcal{P}$ be the property of the analytic function f : *the differential polynomial*

$$F(f)(z) = \bigl(f'(z) - 1\bigr)\bigl(f'(z) - 2\bigr)\bigl(f'(z) - f(z)\bigr)$$

omits the value zero.

(I)   If $\langle f, \mathbb{C} \rangle \in \mathcal{P}$, Picard's first theorem implies $f'(z) = c$ (constant), and so $f(z) = cz + d$. However, $\bigl(f'(z) - f(z)\bigr) \neq 0$, $z \in \mathbb{C}$, means $c = 0$ and $f \equiv$ constant.

(II)   Define

$$\mathcal{F} = \{f_n(z) = nz : z \in U, \ n = 3, 4, 5, \ldots\}.$$

Then $f_n'(z) = n \neq 1, 2$ and $n - nz \neq 0$ for $z \in U$, i.e., each $f_n \in \mathcal{F}$ has the property $\mathcal{P}$. Yet $\mathcal{F}$ is not normal as we have seen before (Example 2.1.3).

Counterexample 4.2.3 Let $\mathcal{P}$ be the property of the meromorphic function f : $f'(z) - af(z)^4 \neq b$, for $a \neq 0$, $b \neq 0, \infty$. Then Pang [1990] has shown that the family

$$\mathcal{F} : \{f : \langle f, \Omega \rangle \in \mathcal{P}\}$$

is normal in any domain Ω. However, according to Mues [1979], there exists a nonconstant meromorphic function f such that $\langle f, \mathbb{C} \rangle \in \mathcal{P}$. This violates the general Converse to the Bloch Principle. See also the discussion pertaining to Theorem 4.4.19.

4.3 Minda's Formalization

In Minda [preprint], a new formalization of a normal property provides another milieu in which normal families can be studied. This approach may at times be easier to apply than that of Robinson-Zalcman.

Definition 4.3.1 *A property $\mathcal{P}$ of analytic functions is called $\mathcal{M}$-normal if the following conditions are satisfied.*

(i)   *If $\langle f, \Omega \rangle \in \mathcal{P}$ and $\Omega' \subseteq \Omega$, then $\langle f, \Omega' \rangle \in \mathcal{P}$.*

(ii)   *If $\langle f, \Omega \rangle \in \mathcal{P}$ and $\phi(z) = az + b$, $a \neq 0$, then $\langle f \circ \phi, \ \phi^{-1}(\Omega) \rangle \in \mathcal{P}$.*

(iii)   *If $\langle f, \Omega \rangle \in \mathcal{P}$ and $c \in \mathbb{C}$, then $\langle f + c, \Omega \rangle \in \mathcal{P}$.*

(iv)   *Let $\langle f_n, \Omega_n \rangle \in \mathcal{P}$, where $\Omega_1 \subseteq \Omega_2 \subseteq \ldots$ and $\mathbb{C} = \cup_{n=1}^{\infty}\Omega_n$. If $f_n \to f$ uniformly on compact subsets of $\mathbb{C}$, then $\langle f, \mathbb{C} \rangle \in \mathcal{P}$.*

(v)   *$\langle I, \mathbb{C} \rangle \notin \mathcal{P}$, where I is the identity function.*

Note that an $\mathcal{M}$-normal property satisfies conditions (i), (ii), and (iii) of a normal property. As for condition (iv) of Definition 4.1.1 we maintain

Theorem 4.3.2 *If a property $\mathcal{P}$ is $\mathcal{M}$-normal and $\langle f, \mathbb{C} \rangle \in \mathcal{P}$, then $f \equiv$ constant.*

Consequently, *if $\mathcal{P}$ is $\mathcal{M}$-normal, it is normal*, and we have by the Heuristic Principle

Corollary 4.3.3 *If $\mathcal{P}$ is $\mathcal{M}$-normal, then the family*

$$\mathcal{F} = \{ f : \langle f, \Omega \rangle \in \mathcal{P} \}$$

is normal in Ω.

In order to prove Theorem 4.3.2 we require

Lemma 4.3.4 *Let $f(z)$ be an entire function. Then $f(z)$ is not linear (i.e., of the form $f(z) = az + b$, $a, b \in \mathbb{C}$) if and only if there exists a sequence $\{z_n\} \subseteq \mathbb{C}$ and positive numbers $\rho_n \to 0$, and a unimodular constant A such that $f(z_n + \rho_n \zeta) - f(z_n) \to A\zeta$, uniformly on compact subsets of $\mathbb{C}$ as $n \to \infty$.*

Proof. In order to establish the sufficiency, assume that $\rho_n \to 0^+$ and $f(z_n + \rho_n \zeta) - f(z_n) \to A\zeta$, uniformly on compact subsets of $\mathbb{C}$, $|A| = 1$. Assuming f is linear with, say, $f' = a$, then

$$1 = |A| = \lim_{n \to \infty} \rho_n |f'(z_n + \rho_n \zeta)| = |a| \lim_{n \to \infty} \rho_n = 0,$$

and we are forced to conclude that f is not linear.

On the other hand, for each $k \in \mathbb{N}$ choose w_k with $|w_k| \leq k$ and satisfying the condition

$$(k - |w_k|)|f'(w_k)| = M_k = \max_{|w| \leq k} (k - |w|)|f'(w)|.$$

We claim that $M_k/k \to \infty$ as $k \to \infty$. To see this, let us suppose that

$$\lim_{k \to \infty} (M_k/k) = M < \infty.$$

Then for a fixed $z \in \mathbb{C}$ and for all k with $|z| < k$,

$$|f'(z)| \leq \frac{M_k}{k - |z|} = \frac{M_k/k}{1 - (|z|/k)},$$

which implies

$$|f'(z)| \leq \lim_{k \to \infty} \frac{M_k/k}{1 - (|z|/k)} = M.$$

Therefore, f' is bounded in $\mathbb{C}$, hence constant by Liouville's theorem. But this contradicts the assumption that f is not linear, and the assertion $M_k/k \to \infty$ is proved.

Next, let $g_k(\zeta)$ be the entire function given by

$$g_k(\zeta) = f(w_k + r_k\zeta) - f(w_k),$$

where

$$r_k = \frac{k - |w_k|}{M_k} = \frac{1}{|f'(w_k)|}.$$

Note that $0 < r_k \leq \frac{k}{M_k}$, so that $r_k \to 0$. Moreover, $g_k(0) = 0$ and $|g_k'(0)| = r_k|f'(w_k)| = 1$. We further claim that the sequence $\{g_k'\}$ is uniformly bounded on compact sets. To this end let $E \subseteq \mathbb{C}$ be a compact set. Then $E \subseteq D(0; M_k)$ for all $k \geq k_0$ for some k_0 because $M_k \to \infty$. Since $\zeta \to w_k + r_k\zeta$ is a mapping of $D(0; M_k)$ onto $D(w_k; k - |w_k|) \subseteq D(0; k)$, we have for $\zeta \in E$, $k \geq k_0$

$$|g_k'(\zeta)| = r_k|f'(w_k + r_k\zeta)| \leq \frac{r_k M_k}{k - |w_k + r_k\zeta|} \leq \frac{k - |w_k|}{k - |w_k| - r_k|\zeta|}$$

$$= \frac{1}{1 - (|\zeta|/M_k)}. \tag{4.5}$$

But $|\zeta|$ is bounded on E and $M_k \to \infty$, so that (4.5) implies $\{g_k'\}$ is uniformly bounded on E. By virtue of Theorem 2.2.6 (or Marty's theorem (§3.3)), $\{g_k\}$ is a normal family, and hence there is a subsequence $\{g_{k_n}\}$ converging normally to an entire function g in $\mathbb{C}$. Evidently, $g(0) = 0$, $|g'(0)| = 1$, and $|g'(\zeta)| \leq 1$ for all $\zeta \in \mathbb{C}$ by (4.5). Then g' is constant by Liouville's theorem, and $g(\zeta) = A\zeta$, with $|A| = 1$. Taking $z_n = w_{k_n}$ and $\rho_n = r_{k_n}$ completes the proof.

We now turn to the proof of Theorem 4.3.2. It is clear from conditions (ii) and (v) that $\langle \phi, \mathbb{C} \rangle \notin \mathcal{P}$ for any linear function $\phi(z) = az + b$, $a \neq 0$. Assume that $\langle f, \mathbb{C} \rangle \in \mathcal{P}$ and f is nonconstant. As f is not linear, according to the lemma there are sequences $\{z_n\}$ and $\{\rho_n\}$ with $\rho_n \to 0$ such that

$$g_n(\zeta) = f(z_n + \rho_n\zeta) - f(z_n) \to A\zeta = g(\zeta), \qquad |A| = 1,$$

normally in $\mathbb{C}$. From conditions (ii) and (iii) we see that $\langle g_n, \mathbb{C} \rangle \in \mathcal{P}$. Then condition (iv) implies $\langle g, \mathbb{C} \rangle \in \mathcal{P}$, contradicting our assertion that no linear function has property $\mathcal{P}$ on $\mathbb{C}$. We conclude that $f \equiv$ constant as desired.

Corollary 4.3.3 can be applied to give a simple proof of Bloch's theorem. First, a few preliminaries are in order.

Consider a function $w = f(z)$ analytic in a domain Ω, $a \in \Omega$ and $b = f(a)$. If $f'(a) = 0$, set $r(a, f) = 0$; otherwise define $r(a, f)$ to be the maximum

$r > 0$ such that the branch of f^{-1} satisfying $f^{-1}(b) = a$ is analytic in the disk $D(b;r)$, and maps this disk into Ω. An equivalent characterization is that $r(a, f)$ is the radius of the largest disk centred about $b = f(a)$ which is the one-sheeted image of a subdomain of Ω under the mapping $w = f(z)$. The radius of such a (*schlicht*) disk is referred to as the *modulus of f at a*. Define the *modulus of univalence of f in Ω* by

$$r(f) = \sup\{r(a, f) : a \in \Omega\}.$$

Observe that for any $c \in \mathbb{C}$, $r(f + c) = r(f)$, and if $\phi : U \to \Omega$ is a one-to-one conformal mapping, then $r(f \circ \phi) = r(f)$. Furthermore,

Lemma 4.3.5 *Let f_n be analytic in a domain Ω_n, with $\Omega_n \subseteq \Omega_{n+1}$, for $n = 1, 2, 3, \ldots$ and $\Omega = \cup_{n=1}^{\infty} \Omega_n$. If $f_n \to f$ uniformly on compact subsets of Ω, then*

$$r(f) \leq \lim_{n \to \infty} r(f_n).$$

Proof. Choose an arbitrary $a \in \Omega$. We may assume $0 < r = r(a, f) < \infty$, for the case $r = \infty$ is done similarly, and if $r = 0$, there is nothing to prove. Take $0 < \varepsilon < \frac{r}{2}$, and let g be the branch of f^{-1} defined in $D(b;r)$ with $g(b) = a$. Setting $\Delta = g\big(D(b;r - \frac{\varepsilon}{2})\big)$, $\overline{\Delta}$ is a compact subset of Ω, and hence $\overline{\Delta} \subseteq \Omega_n$ for all $n \geq n_0$ for some n_0. Since $f_n \to f$ uniformly on $\overline{\Delta}$ and f is univalent on $\overline{\Delta}$, it follows (by a variation of the Hurwitz proof (§1.4)) that f_n is univalent on $\overline{\Delta}$ for all n sufficiently large, let us say for $n \geq n_0$. Moreover, we may assume $|f_n(z) - f(z)| < \frac{\varepsilon}{2}$ for $n \geq n_0$, $z \in \overline{\Delta}$.

Then for $z \in \partial\Delta$,

$$|f_n(z) - b| \geq |f(z) - b| - |f(z) - f_n(z)| > r - \varepsilon > \frac{r}{2},$$

that is, f_n maps $\partial\Delta$ onto some Jordan curve lying in the exterior of the disk $D(b;r - \varepsilon)$, for $n \geq n_0$. As $|f_n(a) - b| < \frac{\varepsilon}{2} < \frac{r}{2}$, we deduce by the Open Mapping Theorem that f_n maps Δ univalently onto a domain containing the disk $D(b;r - \varepsilon)$. Thus $r(a, f_n) \geq r - \frac{3\varepsilon}{2}$, for all $n \geq n_0$, and so

$$r = r(a, f) \leq \lim_{n \to \infty} r(a, f_n).$$

As $a \in \Omega$ is arbitrary, the conclusion follows.

Also required is

Lemma 4.3.6 *For $M > 0$, define $\langle f, \Omega \rangle \in \mathcal{P}_M$ if f is analytic in Ω and $r(f) \leq M$. Then the family*

$$\mathcal{F}_{\mathcal{P}_M} = \{f : \langle f, \Omega \rangle \in \mathcal{P}_M\}$$

is normal in Ω.

Proof. The conditions (i), (ii), (iii) and (v) of Definition 4.3.1 clearly hold, whereas (iv) follows from the preceding lemma. The conclusion obtains from Corollary 4.3.3.

From the above considerations we can prove

Bloch's Theorem [1925]. *There exists some constant $\mathcal{B} > 0$ such that $r(f) \geq \mathcal{B}$ for all functions f analytic in U with $f(0) = 0$, $|f'(0)| = 1$.*

Proof. Set

$$\mathcal{B} = \inf\{r(f) : f \text{ analytic in } U, \ f(0) = 0, \ |f'(0)| = 1\}.$$

Since the identity function is an admissible function, $\mathcal{B} \leq 1$. Consider a sequence of admissible functions $\{f_n\}$ such that $r(f_n) \leq \mathcal{B} + \frac{1}{n}$, $n = 1, 2, 3, \ldots$. By Lemma 4.3.6, $\{f_n\}$ is normal in U, and therefore a subsequence $\{f_{n_k}\}$ converges normally to f in U. Since $f_{n_k}(0) = 0$ and $|f'_{n_k}(0)| = 1$, we have $f(0) = 0$, $|f'(0)| = 1$; in other words, f is admissible and $\mathcal{B} \leq r(f)$. However, in view of Lemma 4.3.5, $r(f) \leq \mathcal{B}$. Hence $r(f) = \mathcal{B}$, and since f is nonconstant, we conclude $\mathcal{B} > 0$.

This proof reveals that extremal functions exist for $\mathcal{B}$. It can be shown that (cf. Goluzin [1969], pp. 363–368, Minda [1982])

$$0.4330127\ldots = \frac{\sqrt{3}}{4} < \mathcal{B} \leq \frac{\Gamma(\frac{1}{3})\,\Gamma(\frac{11}{12})}{\sqrt{\sqrt{3}+1}\,\Gamma(\frac{1}{4})} = 0.4718617\ldots,$$

where the upper bound has been conjectured by Ahlfors and Grunsky to be the true value of $\mathcal{B}$. The lower bound has recently been improved to $\frac{\sqrt{3}}{4} + 10^{-14} < \mathcal{B}$ by Bonk [1990].

The condition for normality of Lemma 4.3.6 was originally deduced by Bloch [1925] as a direct consequence of the above theorem. In fact, let $z_0 \in \Omega$ and take a disk $D(z_0; \rho) \subseteq \Omega$. Assuming $f'(z_0) \neq 0$, the function

$$g(\zeta) = \frac{f(z_0 + \rho\zeta) - f(z_0)}{\rho f'(z_0)}, \qquad f \in \mathcal{F}_{\mathcal{P}_M},$$

is analytic in U, $g(0) = 0$, $|g'(0)| = 1$, with $r(g) \leq \frac{M}{\rho|f'(z_0)|}$. Then

$$\mathcal{B} \leq \frac{M}{\rho|f'(z_0)|},$$

whence

$$|f'(z_0)| \leq \frac{M}{\rho\mathcal{B}},$$

and this is true even if $f'(z_0) = 0$. Since z_0 is arbitrary, one can see that the family

$$\mathcal{F}'_{\mathcal{P}_M} = \{f' : f \in \mathcal{F}_{\mathcal{P}_M}\}$$

is locally bounded in Ω, and hence $\mathcal{F}_{\mathcal{P}_M}$ is normal by Theorem 2.2.6.

There is a related notion due to Landau [1929], who dispensed with Bloch's criterion of univalence, As above, let

$$\mathcal{T} = \{f : f \text{ analytic in } U : f(0) = 0, \; |f'(0)| = 1\},$$

and denote by $\mathcal{L}_f$ the supremum of the set of radii of the open disks covered by $f(U)$. Then the *Landau constant* is given by

$$\mathcal{L} = \inf\{\mathcal{L}_f : f \in \mathcal{T}\},$$

and satisfies $\frac{1}{2} < \mathcal{L} \leq \frac{\Gamma(\frac{1}{3})\Gamma(\frac{5}{6})}{\Gamma(\frac{1}{6})} = 0.5432588\ldots$. For a comprehensive treatment of both the Bloch and Landau constants, refer Sansone and Gerretsen [1969, Vol. II], pp. 73–92, 668–670, Minda [1982], wherein the bounds for $\mathcal{L}$ cited above may be found. The upper bound has been conjectured to be the true value of $\mathcal{L}$. That $f(U)$ covers a disk of radius at least $\mathcal{L} > 0$ for each $f \in \mathcal{T}$ leads to another proof of Picard's first theorem (cf. Hille [1962, Vol. II], Theorem 17.7.2).

4.4   The Drasin Theory

In this section we principally present the work of Drasin [1969], who in a *tour de force* gave several confirmations of the Bloch Principle. The vehicle used to establish these results is the Nevanlinna theory. Some of the techniques have their origins in the work of Milloux [1940] amongst others.

A new proof of the Fundamental Normality Test is given, as well as a generalization of Theorem 3.2.2 which we have already proved via the Heuristic Principle. Theorems 4.4.18 and 4.4.19 answer questions posed by Hayman [1967] regarding Bloch Principle analogues of two theorems in function theory.

In the sequel, $\mathcal{F}$ will denote a family of analytic or meromorphic functions in the unit disk U, although some of the results have obvious extensions to more general domains.

We begin with a reformulation of Montel's theorem (§2.2) more suited to the present context.

Lemma 4.4.1 *Let $\mathcal{F}$ be a family of analytic functions in U, and suppose that there is an increasing finite-valued function $\sum(r)$, $0 < r < 1$, such that for all $f \in \mathcal{F}$,*

$$m(r, f) \leq \sum(r), \qquad r_1 \leq r < 1.$$

Then $\mathcal{F}$ is normal in U.

This was effectively proved in Example 7, §2.3.

We shall wish to make a routine transformation of U via the Möbius function

$$\phi_\alpha(z) = \frac{z + \alpha}{1 + \overline{\alpha}z}, \qquad |\alpha| < 1.$$

If $f \in \mathcal{F}$, write $f_\alpha(z) = f(\phi_\alpha(z))$. Then

$$f_\alpha(0) = f(\alpha), \quad f_\alpha{}'(0) = f'(\alpha)\phi_\alpha'(0) = (1 - |\alpha|^2)f'(\alpha).$$

This transformation will facilitate various arguments in the sequel. Then to each $f \in \mathcal{F}$, we associate a unique $\alpha = \alpha(f)$, and consider the family

$$\mathcal{F}_\alpha = \{f_\alpha : \alpha = \alpha(f), \ f \in \mathcal{F}\}.$$

In the sequel, we always interpret α to mean $\alpha(f)$. The normality of $\mathcal{F}$ and $\mathcal{F}_\alpha$ are equivalent in the following sense.

Lemma 4.4.2 *If $|\alpha| \leq r_0 < 1$, then $\mathcal{F}$ is normal if and only if $\mathcal{F}_\alpha$ is normal.*

Proof. Suppose that $\mathcal{F}_\alpha$ is normal and take $\{f_n\} \subseteq \mathcal{F}$. Then the associated sequence $\{f_{\alpha_n}\} \subseteq \mathcal{F}_\alpha$ has a subsequence $\{f_{\alpha_{n_k}}\}$ converging normally in U to some f. Also $\{\alpha_{n_k}\}$ has a convergent subsequence, say $\{\alpha_m\}$ with $\alpha_m \to \alpha_0$, $|\alpha_0| \leq r_0$. Furthermore, since $\{\phi_\alpha^{-1}\}$ is a normal family (and indeed compact, as we see), the sequence $\{\phi_{\alpha_m}^{-1}\}$ has a convergent subsequence, say $\{\phi_{\alpha_p}^{-1}\}$ with $\phi_{\alpha_p}^{-1} \to \Phi$ normally in U. Therefore,

$$\phi_{\alpha_p}^{-1}(z) = \frac{z - \alpha_p}{1 - \overline{\alpha}_p z} \to \frac{z - \alpha_0}{1 - \overline{\alpha}_0 z} = \Phi(z), \qquad z \in U,$$

and so $|\Phi(z)| < 1$. We conclude that

$$f_p = f_{\alpha_p} \circ \phi_{\alpha_p}^{-1} \to f \circ \Phi$$

normally in U, and $\mathcal{F}$ is a normal family.

The converse follows similarly.

Inequalities. Let f be meromorphic in U, $|\alpha| \leq r_0 < 1$, $f(\alpha) \neq 0, 1, \infty$, $f'(\alpha) \neq 0$. By the Fundamental Inequality (§1.8) with $q = 2$, $a_1 = 0$, $a_2 = 1$, applied to f_α,

$$m(r, f_\alpha) + m\left(r, \frac{1}{f_\alpha}\right) + m\left(r, \frac{1}{f_\alpha - 1}\right)$$
$$\leq 2\,T(r, f_\alpha) - N_1(r, f_\alpha) + S(r, f_\alpha), \qquad 0 < r < 1, \quad (4.6)$$

where

$$N_1(r, f_\alpha) = N\left(r, \frac{1}{f_{\alpha'}}\right) + 2N(r, f_\alpha) - N(r, f_{\alpha'}),$$

$$S(r, f_\alpha) = 2m\left(r, \frac{f_{\alpha'}}{f_\alpha}\right) + m\left(r, \frac{f_{\alpha'}}{f_\alpha - 1}\right) + \log\left|\frac{1}{f_{\alpha'}(0)}\right| + C.$$

Moreover, by Jensen's formula and the First Fundamental Theorem (of Nevanlinna)

$$T(r, f_\alpha) = m\left(r, \frac{1}{f_\alpha}\right) + N\left(r, \frac{1}{f_\alpha}\right) + \log|f_\alpha(0)|$$

$$= m\left(r, \frac{1}{f_\alpha - 1}\right) + N\left(r, \frac{1}{f_\alpha - 1}\right) + \log|f_\alpha(0) - 1| + \varepsilon^*, \quad (4.7)$$

where $|\varepsilon^*| \leq \log 2$.

Adding

$$N(r, f_\alpha) + N\left(r, \frac{1}{f_\alpha}\right) + N\left(r, \frac{1}{f_\alpha - 1}\right),$$

to both sides of (4.6) and using (4.7), together with the relation $N(r, f_{\alpha'}) - N(r, f_\alpha) = \bar{N}(r, f_\alpha)$, yields our basic inequality

$$T(r, f_\alpha) < \bar{N}(r, f_\alpha) + N\left(r, \frac{1}{f_\alpha}\right) + N\left(r, \frac{1}{f_\alpha - 1}\right) - N\left(r, \frac{1}{f_{\alpha'}}\right)$$

$$+ 2m\left(r, \frac{f_{\alpha'}}{f_\alpha}\right) + m\left(r, \frac{f_{\alpha'}}{f_\alpha - 1}\right) + \log\left|\frac{f(\alpha)[f(\alpha) - 1]}{f_{\alpha'}(0)}\right| + C. \quad (4.8)$$

In order to treat the terms involving logarithmic derivatives, we require an estimate due to Nevanlinna (cf. Hayman [1964], Lemma 2.3), and for subsequent work (and in a simplified form), a generalization due to Hiong [1955a].

Lemma 4.4.3 *Let $f(z)$ be meromorphic in $|z| < \rho \leq \infty$. If $f(0) \neq 0, \infty$, then for $0 < r < R < \rho$,*

(i) $m\left(r, \frac{f'}{f}\right) < 4\log^+ T(R, f) + 6\log\frac{1}{R - r} + \log^+\frac{1}{r} + 5\log^+ R$

$$+ 4\log^+\log^+\left|\frac{1}{f(0)}\right| + C \quad (Nevanlinna\ estimate),$$

(ii) $m\left(r, \frac{f^{(k)}}{f}\right) < C\left\{\log^+ T(R, f) + \log^+\frac{1}{R - r} + \log^+\frac{1}{r}\right.$

$$\left. + \log^+ R + \log^+\log^+\left|\frac{1}{f(0)}\right| + 1\right\} \quad (Hiong\ estimate),$$

where $k \in \mathbb{N}$, *and* C *depends only on* k.

If we take $0 < \delta < r < R < 1$, then the terms $\log^+ \frac{1}{r}$ and $\log^+ R$ may be dropped in (i) and (ii), with $C = C(k, \delta)$; if $\delta' > \delta$, we may take $C(k, \delta') = C(k, \delta)$.

Fundamental Normality Test.   We proceed to give another proof of the FNT (§2.7) as an application of the above theory. To this end, consider the family $\mathcal{F}$ of analytic functions in U which omit the values 0 and 1. Note that we then have

$$N(r, f_\alpha) \equiv N\left(r, \frac{1}{f_\alpha}\right) \equiv N\left(r, \frac{1}{f_\alpha - 1}\right) \equiv N(r, f_\alpha') \equiv 0, \qquad f_\alpha \in \mathcal{F}.$$

Then (4.8) becomes $\left(\text{noting } T(r, f_\alpha) \equiv m(r, f_\alpha)\right)$

$$m(r, f_\alpha) < 2\, m\left(r, \frac{f_\alpha'}{f_\alpha}\right) + m\left(r, \frac{f_\alpha'}{f_\alpha - 1}\right) + \log \left| \frac{f(\alpha)[f(\alpha) - 1]}{f_\alpha'(0)} \right| + C.$$

Applying to this estimate say, part (i) of the preceding lemma with $g = f_\alpha$, and $g = f_\alpha - 1$ gives

$$
\begin{aligned}
m(r, f_\alpha) \;\equiv\; & T(r, f_\alpha) < 12 \log^+ T(R, f_\alpha) + 18 \log \frac{1}{R - r} \\
& + 8 \log^+ \log^+ \left| \frac{1}{f(\alpha)} \right| + \log |f(\alpha)| \\
& + 4 \log^+ \log^+ \left| \frac{1}{f(\alpha) - 1} \right| + \log |f(\alpha) - 1| \\
& + \log \left| \frac{1}{f_\alpha'(0)} \right| + C, \qquad \tfrac{1}{2} < r < R < 1. \qquad (4.9)
\end{aligned}
$$

In order to simplify this last expression, we use the inequality

$$\beta \log x + C \log^+ \log^+ \left(\frac{1}{x}\right) < \beta \log^+ x + C', \qquad 0 < x < \infty, \qquad (4.10)$$

which is readily verified whenever $\beta > 0$, $C > 0$. Then (4.9) reduces to

$$
\begin{aligned}
T(r, f_\alpha) \;<\; & 12 \log^+ T(R, f_\alpha) + 18 \log \frac{1}{R - r} + \log^+ |f(\alpha)| \\
& + \log^+ |f(\alpha) - 1| + \log \left| \frac{1}{f_\alpha'(0)} \right| + C, \qquad \tfrac{1}{2} < r < R < 1.
\end{aligned}
$$

Since $|\alpha| \leq r_0$,

$$\left| \frac{1}{\phi_\alpha'(0)} \right| < K = K(r_0),$$

so that

$$\log \left| \frac{1}{f_\alpha'(0)} \right| < \log \left| \frac{1}{f'(\alpha)} \right| + \log K.$$

Therefore, (4.9) reads

$$T(r, f_\alpha) < 12 \log^+ T(R, f_\alpha) + 18 \log \frac{1}{R-r} + 2 \log^+ |f(\alpha)|$$

$$+ \log \left| \frac{1}{f'(\alpha)} \right| + C_1 + K_1, \tag{4.11}$$

where $\frac{1}{2} < r < R < 1$, $|\alpha| \le r_0 < 1$, $K_1 = K_1(r_0)$.

Terms involving $f(\alpha)$ or $f'(\alpha)$ as in (4.8) or in preceding expressions are referred to as *initial-value terms*. In the Nevanlinna theory approach to normal families, control of the initial-value terms presents singular difficulties, as exemplified by the differing methods adopted by Drasin, Yang, and Schwick (cf. §4.5 where the work of the latter two authors is discussed).

To complete the proof, let $\mathcal{F}_1$ denote an arbitrary sequence of functions in $\mathcal{F}$. Now suppose that the following circumstances hold: There are constants $M < \infty$ and $r_0 < 1$ together with an infinite subfamily $\mathcal{F}_2 \subseteq \mathcal{F}_1$ such that $f \in \mathcal{F}_2$ implies there is an $\alpha = \alpha(f)$, $|\alpha| \le r_0$, satisfying

$$2 \log^+ |f(\alpha)| + \log \left| \frac{1}{f'(\alpha)} \right| < M. \tag{4.12}$$

If this were not the case, then for any $M < \infty$, $r_0 < 1$,

$$2 \log^+ |f(\alpha)| + \log \left| \frac{1}{f'(\alpha)} \right| > M, \qquad |\alpha| \le r_0,$$

except for finitely many $f \in \mathcal{F}_1$. But considering separately $|f(\alpha)| \le 1$ and $|f(\alpha)| > 1$ we have

$$\left| 2 \log^+ |f(\alpha)| + \log \left| \frac{1}{f'(\alpha)} \right| - \log \left(\frac{1 + |f(\alpha)|^2}{|f'(\alpha)|} \right) \right|$$

$$= \left| 2 \log^+ |f(\alpha)| - \log \left(1 + |f(\alpha)|^2 \right) \right| \le \log 2,$$

which means that

$$f^\#(z) = \frac{|f'(z)|}{1 + |f(z)|^2} < 1, \qquad |z| \le r_0,$$

except for finitely many $f \in \mathcal{F}_1$. Thus $\mathcal{F}_1$ is normal by Marty's theorem in this particular instance, and so $\mathcal{F}$ is normal.

Hence we may assume that our supposition above holds and apply the inequality (4.11) to each $f \in \mathcal{F}_1$. We further require a technical growth estimate similar to those given by Bureau [1932] and Hiong [1955b]. The proof is given by Drasin [1969].

Lemma 4.4.4 *Let $V(r)$ and $\gamma(r)$ be continuous nondecreasing functions of r, $r_1 < r < 1$. If for some r_0, $r_1 < r_0 < 1$ and $b > 1$, $c > 0$, $M > 0$,*

$$V(r) < c\log^+ V(R) + b\log\frac{1}{R-r} + \gamma(r) + M, \qquad r_0 < r < R < 1,$$

then

$$V(r) < 2b\log\frac{1}{R-r} + 4\gamma(R) + M_1, \qquad r_0 < r < R < 1,$$

where $M_1 = 4M + 16c^2 + 4c[\log b + 3\log 2] = M_1(M, b, c)$.

Note that is is possible to make the conclusion of Lemma 4.4.4 independent of R by taking $R = \frac{r+1}{2}$.

Setting $V(r) = T(r, f_\alpha)$ and $\gamma(r) \equiv 0$ in the lemma, then $m(r, f_\alpha) \equiv T(r, f_\alpha)$ is uniformly bounded by an increasing function of r for each $f \in \mathcal{F}_2$, $\frac{1}{2} < r < 1$ (invoking (4.11) and (4.12) and the lemma). The normality of $\mathcal{F}$ then follows from Lemmas 4.4.1 and 4.4.2 , and the proof of the FNT is complete.

A Generalization. In this section we prove a generalization of Montel's theorem 3.2.2 which has already been considered within the context of the Heuristic Principle (cf. Theorem 4.1.3). The proof from the point of view of the Nevanlinna theory is very instructive.

Theorem 4.4.5 *Let $\mathcal{F}$ be a family of meromorphic functions in U. If, for each $f \in \mathcal{F}$, the poles are of multiplicity $\geq h$, the zeros are of multiplicity $\geq k$, and the zeros of $f(z) - 1$ are of multiplicity $\geq \ell$, with*

$$\frac{1}{h} + \frac{1}{k} + \frac{1}{\ell} = \mu < 1,$$

then $\mathcal{F}$ is normal.

Proof. Again we take $r_0 < 1$, $|\alpha| \leq r_0$ with $f(\alpha) \neq 0, 1, \infty$, $f'(\alpha) \neq 0$. In this case the hypotheses imply

$$\bar{N}(r, f_\alpha) \leq \frac{1}{h}N(r, f_\alpha) \leq \frac{1}{h}T(r, f_\alpha),$$

$$\bar{N}\left(r, \frac{1}{f_\alpha}\right) \leq \frac{1}{k}N\left(r, \frac{1}{f_\alpha}\right) \leq \frac{1}{k}T\left(r, \frac{1}{f_\alpha}\right) = \frac{1}{k}(T(r, f_\alpha) - \log|f(\alpha)|),$$

$$\bar{N}\left(\frac{1}{f_\alpha - 1}\right) \leq \frac{1}{\ell}N\left(\frac{1}{f_\alpha - 1}\right) \leq \frac{1}{\ell}T\left(\frac{1}{f_\alpha - 1}\right)$$

$$\leq \frac{1}{\ell}(T(r, f_\alpha) - \log|f(\alpha) - 1|) + \log 2,$$

the latter inequality resulting from the First Fundamental Theorem. Then the basic inequality (4.8) becomes

$$T(r, f_\alpha) < \mu T(r, f_\alpha) + 2m\left(r, \frac{f_\alpha{}'}{f_\alpha}\right) + m\left(r, \frac{f_\alpha{}'}{f_\alpha - 1}\right)$$

$$+ \left(1 - \frac{1}{k}\right)\log|f(\alpha)| + \left(1 - \frac{1}{\ell}\right)\log|f(\alpha) - 1| + \log\left|\frac{1}{f_\alpha{}'(0)}\right| + C.$$

We now bring Lemma 4.4.3 to bear on the logarithmic derivative terms as before, giving

$$(1 - \mu)T(r, f_\alpha) < 12\log^+ T(R, f_\alpha) + 18\log\frac{1}{R - r} + 8\log^+\log^+\left|\frac{1}{f(\alpha)}\right|$$

$$+ \left(1 - \frac{1}{k}\right)\log|f(\alpha)| + 4\log^+\log^+\left|\frac{1}{f(\alpha) - 1}\right|$$

$$+ \left(1 - \frac{1}{\ell}\right)\log|f(\alpha) - 1| + \log\left|\frac{1}{f_\alpha{}'(0)}\right| + C$$

$$< 12\log^+ T(R, f_\alpha) + 18\log\frac{1}{R - r}$$

$$+ \left(2 - \frac{1}{k} - \frac{1}{\ell}\right)\log^+|f(\alpha)| + \log\left|\frac{1}{f'(\alpha)}\right| + C, \quad (4.13)$$

$\frac{1}{2} < r < R < 1$, where (4.10) has been used to obtain this last expression.

We demonstrate that $\mathcal{F}$ is normal in a neighbourhood of each point $z \in U$, for which it suffices to consider only a neighbourhood of $z = 0$. Letting $\mathcal{F}_1$ be a sequence of functions in $\mathcal{F}$, we consider two cases.

Case 1. Suppose that there are functions $f_n \in \mathcal{F}_1$, $n = 1, 2, 3, \ldots$, numbers $\alpha_n = \alpha(f_n) \neq 0$, $\alpha_n \to 0$, $f_n(\alpha_n) \neq 0, 1, \infty$, $f_n'(\alpha_n) \neq 0$, and $M < \infty$ such that

$$\left(2 - \frac{1}{k} - \frac{1}{\ell}\right)\log^+|f_n(\alpha_n)| + \log\left|\frac{1}{f_n'(\alpha_n)}\right| < M.$$

This condition, together with Lemma 4.4.4, applied to (4.13) yields

$$T(r, f_{n, \alpha_n}) < \sum(r), \qquad \frac{1}{2} < r < 1,$$

where $f_{n, \alpha_n} = f_n \circ \phi_{\alpha_n}$ and $\sum(r)$ is an increasing finite-valued function. Then $\{f_{n, \alpha_n}\}$ is normal in a neighbourhood of the origin by virtue of Theorem 3.6.9, and likewise for $\{f_n\}$, and hence $\mathcal{F}_1$.

Case 2. For some $\delta > 0$ such that if $|\alpha| < \delta$ and $f(\alpha) \neq 0, 1, \infty$, $f'(\alpha) \neq 0$, we have

$$\left(2 - \frac{1}{k} - \frac{1}{\ell}\right)\log^+|f(\alpha)| + \log\left|\frac{1}{f'(\alpha)}\right| > M,$$

except for finitely many $f \in \mathcal{F}_1$. From this inequality, just as in the preceding theorem, we can show that $f^{\#}(\alpha) \leq 1$ for $|\alpha| < \delta$, $f(\alpha) \neq 0, 1, \infty$, $f'(\alpha) \neq 0$. By continuity, $f^{\#}(z) \leq 1$ in $|z| < \delta$, so that $\mathcal{F}_1$ is normal in this neighbourhood, completing the proof of the theorem.

A New Composition. We next require another composition akin to f_α in order to establish some further inequalities required in the sequel.

Let $r_0 < 1$ be fixed and $r_0 < r < 1$. For $|\alpha| \leq r_0$, define

$$\psi_\alpha(z) = r^2 \frac{z - \alpha}{r^2 - \bar{\alpha} z};$$

ψ_α is just the one-to-one conformal mapping of $|z| \leq r$ onto itself. Although ψ_α depends on r, the specific choice of r will be obvious from the context. A slight complication arises from the fact that the composition $f \circ \psi_\alpha$ is not defined in all of U. However, it is defined if $|z| < r^2(1 + r_0)/(r_0 + r^2)$.

Lemma 4.4.6 *If g is meromorphic in $|z| \leq r$, and $|\alpha| \leq r_0 < r$, then there is a positive constant $K = K(r_0, r)$ such that*

$$\frac{1}{K} m(r, g) \leq m(r, g \circ \psi_\alpha) \leq K m(r, g).$$

Proof. If $\psi_\alpha(r e^{i\theta}) = r e^{i\tau}$, $0 \leq \theta \leq 2\pi$, $|\alpha| \leq r_0 < r$, then, by the change of variable formula,

$$
\begin{aligned}
m(r, g) &= \frac{1}{2\pi} \int_0^{2\pi} \log^+ |g(r e^{i\tau})| \, d\tau \\
&= \frac{1}{2\pi} \int_0^{2\pi} \log^+ |g \circ \psi_\alpha(r e^{i\theta})| \frac{r^2 - |\alpha|^2}{|r - \bar{\alpha} e^{i\theta}|^2} \, d\theta \\
&\leq \frac{r + |\alpha|}{r - |\alpha|} m(r, g \circ \psi_\alpha) \leq \frac{r + r_0}{r - r_0} m(r, g \circ \psi_\alpha) = K m(r, g \circ \psi_\alpha).
\end{aligned}
$$

Similarly for the reverse inequality.

Remark 4.4.7 It is important to note that if $s > r$, we may take $K(r_0, s) = K(r_0, r)$; moreover, K may be taken arbitrarily close to 1 provided r/r_0 is sufficiently large.

A typical line of reasoning used in the sequel is as follows: *If $\mathcal{F}$ is not normal in U, then there exists some $r_0 < 1$ and an infinite subfamily $\mathcal{F}_1 \subseteq \mathcal{F}$ such that for each $f \in \mathcal{F}_1$ there corresponds some $\alpha = \alpha(f)$, $|\alpha| \leq r_0$ with*

$$|f(\alpha)| \geq 1. \tag{4.14}$$

In fact, if this were not the case, then, for each $r < 1$, (4.14) holds only for finitely many $f \in \mathcal{F}$, i.e., $|f(z)| < 1$ for $|z| < r$ for all $f \in \mathcal{F}_1(r) \subseteq \mathcal{F}$, where $\mathcal{F} - \mathcal{F}_1(r)$ is finite. Then $\mathcal{F}_1(r)$ is normal in $|z| < r < 1$, implying $\mathcal{F}$ is normal there. However, as $\mathcal{F}$ is not normal in U, $\mathcal{F}$ is not normal in some $|z| < r_0 < 1$, and this contradiction establishes the assertion.

In the same vein we have that: *If $\mathcal{F}$ is not normal at $z = 0$, then there exists a sequence $\{f_n\} \subseteq \mathcal{F}$ and points $\alpha_n = \alpha_n(f_n)$ with $\alpha_n \to 0$ as $n \to \infty$, such that $|f_n(\alpha_n)| \geq 1$.*

We proceed to establish some further lemmas by making use of the new composition $f \circ \psi_\alpha$. The first is a variant of Lemma 4.4.3 which has the virtue of dispensing with the initial-value terms.

Lemma 4.4.8 *Let $\mathcal{F}$ be a nonnormal family of analytic functions in U. Then there exits $r_0 < 1$ such that*

$$m\left(r, \frac{f'}{f}\right) < A + B \log^+ T(R, f) + C \log \frac{1}{R - r}, \qquad \tfrac{1}{2}(1 + r_0) < r < R < 1,$$

for all f belonging to an infinite subfamily $\mathcal{F}_1 \subseteq \mathcal{F}$.

Proof. Since $\mathcal{F}$ is not normal, there is an $r_0 < 1$ such that there exists an $\alpha = \alpha(f)$, $|\alpha| \leq r_0$, satisfying $|f(\alpha)| \geq 1$ for all f belonging to an infinite subfamily $\mathcal{F}_1 \subseteq \mathcal{F}$. Applying the preceding lemma and then Lemma 4.4.3 with $g = f \circ \psi_{-\alpha}$, $f \in \mathcal{F}_1$, so that $g(0) = f(\alpha) \neq 0, \infty$, we have

$$\begin{aligned}
m\left(r, \frac{f'}{f}\right) &\leq Km\left(r, \frac{f'}{f} \circ \psi_{-\alpha}\right) \\
&< K\left(4 \log^+ T(r', f \circ \psi_{-\alpha}) + 6 \log \frac{1}{r' - r} + D\right),
\end{aligned}$$

$$\tfrac{1}{2}(1 + r_0) < r < r' < \frac{r^2(1 + r_0)}{(r_0 + r^2)}. \tag{4.15}$$

Now the circle $|\psi_{-\alpha}(\zeta)| = r'$ is contained in the circle

$$|z| = \frac{r^2(r' - r_0)}{r^2 - r_0 r'} = r_1, \tag{4.16}$$

and f analytic implies by Theorem 1.8.2 and the Maximum Modulus Theorem that

$$\begin{aligned}
T(r', f \circ \psi_{-\alpha}) &\leq \log^+ M(r', f \circ \psi_{-\alpha}) \leq \log^+ M(r_1, f) \\
&\leq \frac{R + r_1}{R - r_1} T(R, f), \qquad r_1 < R < 1.
\end{aligned}$$

Then

$$\log^+ T(r', f \circ \psi_{-\alpha}) \leq \log \frac{1}{R - r_1} + \log^+ T(R, f) + A_1, \tag{4.17}$$

and A_1 is independent of f and α. Given $R > r$, it is possible to find r' satisfying

$$\frac{1}{r' - r} = \frac{1}{R - r_1},$$

by considering the graph of r_1 as given by (4.16). This permits the log terms of (4.15) and (4.17) to be combined. We can then find a constant $E = E(r_0)$ (based on an analysis of (4.16) when $r_1 = R$) such that $\frac{1}{R-r_1} \leq \frac{E}{R-r}$, from which the conclusion follows.

By considering the family $\{e^{n(z-1)} : |z| < 1,\ n = 1, 2, 3, \ldots\}$, it can be verified that the nonnormality of $\mathcal{F}$ is required.

We restate Lemma 4.4.8 in a local version, omitting the routine proof.

Lemma 4.4.9 *Let $\mathcal{F}$ be a family of analytic functions in $|z| < \delta$ $(\delta > 0)$ which is not normal in any neighbourhood of the origin. Then there exist constants $A = A(\delta)$, $B = B(\delta)$, $C = (\delta)$ such that*

$$m\left(r, \frac{f'}{f}\right) < A + B \log^+ T(R, f) + C \log \frac{1}{R-r}, \qquad \frac{\delta}{2} < r < R < \delta,$$

for all f belonging to an infinite subfamily of $\mathcal{F}$.

Digressing for a moment, Schwick [1990] has shown that the estimate of Lemma 4.4.9 does not hold for families of meromorphic functions. Indeed, consider the sequence

$$f_n(z) = \frac{e^{n(z-1)}}{z + \frac{1}{n}}, \qquad n = 1, 2, 3, \ldots .$$

Let $z_n = -\frac{1}{n}$ and $\rho_n = e^{-n}$. Then

$$f_n(z_n + \rho_n z) = \frac{e^{ne^{-n}z}}{ez} \rightarrow \frac{1}{ez}$$

spherically uniformly on compact subsets of $\mathbb{C}$. By the Zalcman Lemma (§4.1 and §4.5), $\{f_n\}$ is not a normal family at $z = 0$. Moreover, $m\left(r, \frac{f'_n}{f_n}\right)$ increases as $\log^+ n$, but the right-hand side of the Lemma 4.4.9 estimate is $O(\log^+ \log^+ n)$. This proves the assertion.

On the other hand, Schwick demonstrates that the following estimate does hold.

Theorem 4.4.10 *Let $\mathcal{F}$ be a family of meromorphic functions in U which is not normal at $z = 0$, and $f(0) \neq \infty$ for $f \in \mathcal{F}$. Then there exists a sequence $\{f_n\} \subseteq \mathcal{F}$, such that*

$$m\left(r, \frac{f'_n}{f_n}\right) < A \log^+ T(R, f_n) + B \log \frac{1}{R-r} + C \log^+ \log^+ L_n + D,$$

for $\frac{1}{2} < r < R < 1$, and $L_n = \max\limits_{|z| \leq \frac{1}{n}} \frac{|f'_n(z)|}{1 + |f_n(z)|^2}$.

Applying the techniques of Lemmas 4.4.8 and 4.4.9 we obtain another useful result concerning the proximity function.

Lemma 4.4.11 *Let $\mathcal{F}$ be a family of analytic functions in U (resp. $|z| < \delta$) which is not normal (resp. not normal at the origin). Then for some $r_0 < 1$ and constant K,*

$$m\left(r, \frac{1}{f}\right) \leq Km(r, f), \qquad r_0 < r < 1 \ \left(resp. \ \tfrac{\delta}{2} < r < \delta\right),$$

for all f belonging to an infinite subfamily $\mathcal{F}_1$ of $\mathcal{F}$.

Proof. As before, there is an infinite subfamily $\mathcal{F}_1$ of $\mathcal{F}$ such that for some $r_0 < 1$, $|f(\alpha)| \geq 1$ at $\alpha = \alpha(f)$, $|\alpha| \leq r_0$, $f \in \mathcal{F}_1$ (resp. there is a sequence $\alpha_n \to 0$ such that $|\alpha_n| < \tfrac{\delta}{3}$, say, $n = 1, 2, 3, \ldots$ and $|f_n(\alpha_n)| \geq 1$, $f_n \in \mathcal{F}_1$). If $r > r_0$ (resp. $r > \tfrac{\delta}{2}$), then Lemma 4.4.6 implies, for $f \in \mathcal{F}_1$,

$$
\begin{aligned}
m\left(r, \frac{1}{f}\right) &\leq Km\left(r, \frac{1}{f} \circ \psi_{-\alpha}\right) \leq K \cdot T\left(r, \frac{1}{f} \circ \psi_{-\alpha}\right) \\
&\leq K\big(T(r, f \circ \psi_{-\alpha}) - \log|f(\alpha)|\big) \\
&\leq K \cdot T(r, f \circ \psi_{-\alpha}) = Km(r, f \circ \psi_{-\alpha}) \leq K^2 m(r, f).
\end{aligned}
$$

There is an important extension of Lemma 4.4.8 which we now endeavour to prove.

Lemma 4.4.12 *Suppose that $\mathcal{F}$ is a family of analytic functions in U (resp. $|z| < \delta$) which is not normal at the origin, and fix $k \in \mathbb{N}$. Then there are positive constants A, B, C, (resp. $A = A(\delta)$, $B = B(\delta)$, $C = C(\delta)$) such that*

$$m\left(r, \frac{f^{(k)}}{f}\right) < A + B\log^+ T(R, f) + C\log\frac{1}{R - r}, \qquad \tfrac{1}{2} < r < R < 1,$$
$$\left(resp. \ \tfrac{\delta}{2} < r < R < \delta\right),$$

for all f belonging to a countably infinite subfamily $\mathcal{F}_1 \subseteq \mathcal{F}$.

The origins of this result lie in Theorem 1 of Hayman [1953], and a slight variation of the present lemma was established in Drasin [1969]. The proof given here, by induction, is due to Schwick [1989] and gives us a glimpse of the power of the Zalcman Lemma in dealing with the initial-value terms. (In §4.5 this technique is exploited further). In fact, the case $k = 1$ is just Lemma 4.4.9. Let us assume then that the result is proved for integers $1, \ldots, k - 1$. Employing a local version of the Zalcman Lemma (refer to §4.5), there exists $z_n \to 0$, $\rho_n \to 0$, functions $f_n \in \mathcal{F}$, and a nonconstant entire function $g(z)$ such that $f_n(z_n + \rho_n z) \to g(z)$ normally in $\mathbb{C}$. Take $z_0 \in \mathbb{C}$ satisfying $g(z_0) \neq 0 \ \infty$, and, for $\tfrac{1}{2} < r < 1$, let

$$\psi_n(z) = r^2 \frac{z + (z_n + \rho_n z_0)}{r^2 + (z_n + \rho_n z_0)z}.$$

In view of Lemma 4.4.6 there exists a positive constant K such that for any meromorphic function h and all n sufficiently large (and hence without loss of generality for all $n \in \mathbb{N}$)

$$\frac{1}{K}m(r,h) \le m(r, h \circ \psi_n) \le Km(r,h),$$

for $\frac{1}{2} < r < 1$. We shall also require the identity

$$(f_n \circ \psi_n)^{(k)} = (f_n^{(k)} \circ \psi_n)(\psi_n')^k + \sum_{i=1}^{k-1}(f_n^{(i)} \circ \psi_n)P_i(\psi_n),$$

where P_i is a differential polynomial. As a consequence,

$$m\left(r, \frac{f_n^{(k)}}{f_n}\right) \le O(1)\, m\left(r, \frac{f_n^{(k)}}{f_n} \circ \psi_n\right)$$

$$\le O(1)\left\{ m\left(r, \left(\frac{f_n^{(k)}}{f_n} \circ \psi_n\right)(\psi_n')^k\right) + m\left(r, \frac{1}{(\psi_n')^k}\right)\right\}$$

$$\le O(1)\left\{ m\left(r, \frac{(f_n \circ \psi_n)^{(k)}}{(f_n \circ \psi_n)}\right) + \sum_{i=1}^{k-1} m\left(r, \frac{f_n^{(i)}}{f_n} \circ \psi_n\right)\right.$$

$$\left. + \sum_{i=1}^{k-1} m(r, P_i(\psi_n)) + 1\right\}$$

$$\le O(1)\left\{ m\left(r, \frac{(f_n \circ \psi_n)^{(k)}}{(f_n \circ \psi_n)}\right) + \sum_{i=1}^{k-1} m\left(r, \frac{f_n^{(i)}}{f_n}\right) + 1\right\}, \qquad \tfrac{1}{2} < r < 1.$$

The induction hypothesis is now applied to the summation on the right. An application of the Hiong estimate (Lemma 4.4.3) to the preceding term gives

$$m\left(r, \frac{(f_n \circ \psi_n)^{(k)}}{(f_n \circ \psi_n)}\right) < O(1)\left\{ \log^+ T(r', f_n \circ \psi_n) + \log \frac{1}{r' - r}\right.$$

$$\left. + \log^+ \log^+ \frac{1}{|f_n(z_n + \rho_n z_0)|} + 1\right\}$$

$$< O(1)\left\{ \log^+ T(r', f_n \circ \psi_n) + \log \frac{1}{r' - r} + 1\right\},$$

$\frac{1}{2} < r < r' < R < 1$. The proof is then completed along the lines of Lemma 4.4.8.

As a consequence of the above lemma, a comparison of $T(r, f^{(k)})$ and $T(r, f)$ for a nonnormal family can be made (cf. also Lemma 4.5.14).

Corollary 4.4.13 *Let $\mathcal{F}$ be a family of analytic functions in U (resp. $|z| < \delta$), which is not normal at the origin, and fix $k \in \mathbb{N}$. Then there are* positive constants A, B, C, (resp. $A = A(\delta)$, $B = B(\delta)$, $C = C(\delta)$) and a countably infinite subfamily $\mathcal{F}_1 \subseteq \mathcal{F}$ with

$$T(r, f^{(k)}) < A + B \cdot T(R, f) + C \log \frac{1}{R - r}, \qquad \tfrac{1}{2} < r < R < 1,$$
$$\left(resp. \ \tfrac{\delta}{2} < r < R < \delta\right),$$

for all $f \in \mathcal{F}_1$.

Proof. Using a simple estimate and the lemma,

$$
\begin{aligned}
T(r, f^{(k)}) &\leq m\left(r, \frac{f^{(k)}}{f}\right) + T(r, f) \\
&< A + B \log^+ T(R, f) + C \log \frac{1}{R - r} + T(r, f) \\
&< A + B \cdot T(R, f) + C \log \frac{1}{R - r},
\end{aligned}
$$

as desired.

Zero-Free Families of Analytic Functions. It is well-known that if $f(z)$ is entire, $f(z) \neq 0$, $f'(z) \neq 1$, then $f(z)$ reduces to a constant. In a lecture in 1934 (*Le rôle des familles normales*), Montel conjectured that a family of analytic functions in a domain where $f(z) \neq 0$, $f'(z) \neq 1$, would constitute a normal family, but did not give a proof. This was given a year later by Miranda [1935], who, using the Nevanlinna theory, was able to replace the derivative condition by: $f^{(k)}(z) \neq 1$ for a given $k \in \mathbb{N}$ (cf. also Hiong [1958, 1967]). The meromorphic case, and related results, are treated in §4.5.

We prove a generalization of the Montel-Miranda theorem, namely, one due to Chuang [1940].

Theorem 4.4.14 *Let $\mathcal{F}$ be a family of zero-free analytic functions in U, and $a_0(z), \ldots, a_{k-1}(z)$ fixed analytic functions. Suppose that $\mathcal{G}$ consists of the functions given by*

$$g(z) = f^{(k)}(z) + a_{k-1}(z) f^{(k-1)}(z) + \ldots + a_0(z) f(z), \qquad f \in \mathcal{F}, \quad (4.18)$$

and that $g(z) \neq 1$ for all $z \in U$, $g \in \mathcal{G}$. Then $\mathcal{F}$ is normal in U.

The proof given here is due to Drasin; the generality of some of the reasoning will permit their use in subsequent arguments. Throughout, each subfamily $\mathcal{F}_i \subseteq \mathcal{F}$ will have a corresponding subfamily $\mathcal{G}_i \subseteq \mathcal{G}$, and likewise g_α and f_α will be related by (4.18).

Let $\mathcal{F}_1$ be a countably infinite subfamily of $\mathcal{F}$. We will prove that $\mathcal{F}$ is normal at an arbitrary point $z_0 \in U$. To this end, take an open disk

$D(z_0; \delta) \subseteq U$, and without loss of generality set $z_0 = 0$. For the present, we shall assume the validity of the following condition:

If $\mathcal{G}$ is normal at the origin, then $\mathcal{F}$ is normal at the origin.   (4.19)

We may then presume the corresponding subfamily $\mathcal{G}_1 \subseteq \mathcal{G}$ is not normal at the origin, for otherwise by (4.19) there is nothing to prove. Thus, in view of Marty's theorem (§3.3), there is an infinite subfamily $\mathcal{G}_2 \subseteq \mathcal{F}_1$ for which there exists an $r_0 < \delta$ and points $\alpha = \alpha(g)$, $|\alpha| \leq r_0$ such that

$$\frac{|g'(\alpha)|}{1 + |g(\alpha)|^2} > 2, \qquad g \in \mathcal{G}_2.$$

It may be assumed that $g(\alpha) \neq 0$, $g'(\alpha) \neq 0$, by considering a point arbitrarily close to α if need be. Then for $g \in \mathcal{G}_2$, this implies

$$\frac{|g(\alpha) - 1|}{|g'(\alpha)|} < 1. \tag{4.20}$$

From the definition of g_α we have

$$\log^+ T(R', g_\alpha) \leq \sum_{j=0}^{k-1} \log^+ T(R', a_{j,\alpha}) + \sum_{j=0}^{k} \log^+ T(R', f^{(j)}{}_\alpha) + D$$

$$= L_\alpha(R') + \sum_{j=0}^{k} \log^+ T(R', f^{(j)}{}_\alpha) + D, \tag{4.21}$$

having set $L_\alpha(R') = \sum_{j=0}^{k-1} \log^+ T(R', a_{j,\alpha})$, and D is an absolute constant.

Supposing $\mathcal{G}_2$ to have no convergent subfamily, apply Corollary 4.4.13 to (4.21). The conclusion is that there are infinite subfamilies $\mathcal{G}_3$ of $\mathcal{G}_2$ and $\mathcal{F}_3$ of $\mathcal{F}_2$ whose members satisfy

$$\log^+ T(R', g_\alpha) \leq L_\alpha(R') + A \log \frac{1}{R - R'} + B \log^+ T(R, f_\alpha) + C,$$
$$\tfrac{\delta}{2} < R' < R < \delta.$$

Next, apply Lemma 4.4.12 and the preceding estimate to the basic inequality (4.8) with g_α replacing f_α, $R' = \frac{1}{2}(r + R)$. Note that $g(\alpha) \neq 0, 1, \infty$, $g'(\alpha) \neq 0$ for $g \in \mathcal{G}_3$. Then

$$T(r, g_\alpha) < N\left(r, \frac{1}{g_\alpha}\right) + N\left(r, \frac{1}{g_\alpha - 1}\right) - N\left(r, \frac{1}{g_\alpha'}\right) + A \log^+ T(R, f_\alpha)$$

$$+ \; B \log \frac{1}{R - r} + C \cdot L_\alpha(R) + \log\left|\frac{g(\alpha)[g(\alpha) - 1]}{g'(\alpha)}\right| + D, \tag{4.22}$$

where $\frac{\delta}{2} < r < R < \delta$, and g belongs to an infinite subfamily $\mathcal{G}_4 \subseteq \mathcal{G}_3$. Moreover, if $|\alpha| \leq r_0 < 1$, then $L_\alpha(R) < K$, where $K = K(r_0, \delta)$.

We next wish to eliminate the g_α terms from (4.22). Using the fact that

$$m\left(r,\frac{1}{g_\alpha}\right) + N\left(r,\frac{1}{g_\alpha}\right) = T(r,g_\alpha) - \log|g(\alpha)|,$$

(4.22) reduces to

$$m\left(r,\frac{1}{g_\alpha}\right) < N\left(r,\frac{1}{g_\alpha-1}\right) - N\left(r,\frac{1}{g_{\alpha'}}\right) + A\log^+ T(R,f_\alpha)$$
$$+ B\log\frac{1}{R-r} + \log\left|\frac{g(\alpha)-1}{g'(\alpha)}\right| + C,$$
$$\tfrac{\delta}{2} < r < R < \delta. \quad (4.23)$$

Furthermore,

$$T\left(r,\frac{1}{f_\alpha}\right) = m\left(r,\frac{1}{f_\alpha}\right) + N\left(r,\frac{1}{f_\alpha}\right)$$
$$\leq m\left(r,\frac{g_\alpha}{f_\alpha}\right) + m\left(r,\frac{1}{g_\alpha}\right) + N\left(r,\frac{1}{f_\alpha}\right). \quad (4.24)$$

According to Lemmas 4.4.6 and 4.4.12 there are two infinite subfamilies $\mathcal{F}_5 \subseteq \mathcal{F}_4$, $\mathcal{G}_5 \subseteq \mathcal{G}_4$ with

$$m\left(r,\frac{g_\alpha}{f_\alpha}\right) \leq \sum_{j=0}^{k-1} m(r,a_{j,\alpha}) + \sum_{j=1}^{k} m\left(r,\frac{f^{(j)}}{f}\circ\phi_\alpha\right) + C$$
$$\leq H_\alpha(r) + A\log^+ T(R,f_\alpha) + B\log\frac{1}{R-r} + C, \quad (4.25)$$

for $\tfrac{\delta}{2} < r < R < \delta$. Here $H_\alpha(r) = \sum_{j=0}^{k-1} m(r,a_{j,\alpha})$ and for $|\alpha| \leq r_0$, $H_\alpha(r) < H = H(r_0,\delta)$ as before. Putting (4.25) and (4.23) in (4.24) gives

$$T\left(r,\frac{1}{f_\alpha}\right) \leq N\left(r,\frac{1}{f_\alpha}\right) + N\left(r,\frac{1}{g_\alpha-1}\right) - N\left(r,\frac{1}{g_{\alpha'}}\right) + A\log^+ T(R,f_\alpha)$$
$$+ B\log\frac{1}{R-r} + \log\left|\frac{g(\alpha)-1}{g'(\alpha)}\right| + C,$$
$$\tfrac{\delta}{2} < r < R < \delta. \quad (4.26)$$

As we have assumed, $|\alpha| \leq r_0$, A, B and C depend only on δ and the family $\mathcal{F}_2$. According to the hypotheses, (4.26) simplifies to

$$T\left(r,\frac{1}{f_\alpha}\right) \leq A\log^+ T(R,f_\alpha) + B\log\frac{1}{R-r}$$
$$+ \log\left|\frac{g(\alpha)-1}{g'(\alpha)}\right| + C, \quad \tfrac{\delta}{2} < r < R < \delta. \quad (4.27)$$

Lastly, since functions in $\mathcal{F}$ are zero-free, analytic, unless $\mathcal{F}_5$ is normal, the local form of Lemma 4.4.11 may be applied to yield

$$T(r, f_\alpha) \le K \cdot T\left(r, \frac{1}{f_\alpha}\right), \tag{4.28}$$

for $\frac{\delta}{2} < r < \delta$, with f belonging to an infinite subfamily $\mathcal{F}_6 \subseteq \mathcal{F}_5$.

Combining (4.28) and (4.27), with α and r_0 in accordance with (4.20), we next apply Lemma 4.4.4 with $V(r) = T(r\delta, f_\alpha)$, and $\gamma(r) \equiv 0$, $f \in \mathcal{F}_6$. With a subsequent application of Lemma 4.4.1, we conclude that $\mathcal{F}$ is normal in $|z| < \delta$, as desired.

Regarding condition (4.19), suppose that $\{g_n\} \subseteq \mathcal{G}$ and that $\{g_n\}$ converges uniformly on compact subsets of $|z| < \delta$ to g_0. We claim that the corresponding sequence $\{f_n\} \subseteq \mathcal{F}$ is normal at the origin.

Case 1. $g_0(0) \neq 0$. Then there is a neighbourhood $|z| < \eta < \delta$ on which $|g(z)| > \varepsilon > 0$ for g belonging to an infinite subfamily $\mathcal{G}_1 \subseteq \{g_n\}$. We may assume that $\frac{\delta}{2} < \eta < \delta$. Then by virtue of (4.25), unless $\mathcal{F}_1$ is normal at the origin, there exists $\mathcal{F}_2 \subseteq \mathcal{F}_1$ such that

$$
\begin{aligned}
T\left(r, \frac{1}{f}\right) = m\left(r, \frac{1}{f}\right) &\le m\left(r, \frac{g}{f}\right) + m\left(r, \frac{1}{g}\right) \\
&\le H_0(r) + A\log^+ T(R, f) + B\log\frac{1}{R-r} + C \\
&\le A\log^+ T(R, f) + B\log\frac{1}{R-r} + C + H_0(\delta),
\end{aligned}
$$

$\frac{\delta}{2} < r < R < \eta < \delta$, $f \in \mathcal{F}_2$, where we have used the fact that $m\left(r, \frac{1}{g}\right) = O(1)$. An application of Lemmas 4.4.11 and 4.4.4 with $V(r) = T(r\eta, f)$, $\gamma(r) \equiv 0$, yields the normality of $\mathcal{F}_1$ as in the above argument.

Case 2. $g_0(0) = 0$. By the method of variation of parameters (cf. Coddington and Levinson [1955], p. 87), we can solve for f_n in the equation

$$g_n(z) = f_n^{(k)}(z) + a_{k-1}(z)f_n^{(k-1)}(z) + \ldots + a_0(z)f_n(z)$$

to give

$$f_n(z) = \sum_{m=1}^{k} \left(\alpha_{n,m} + \beta_{n,m}(z)\right) h_m(z), \tag{4.29}$$

where $h_1(z), \ldots, h_k(z)$ are linearly independent solutions of the homogeneous equation $g(z) = 0$, for g given by (4.18), the $\alpha_{n,m}$ are arbitrary constants, and

$$\beta_{n,m}(z) = \int_0^z \frac{W_m(h_1, \ldots, h_k)(\zeta)}{W(h_1, \ldots, h_k)(\zeta)} g_n(\zeta)\, d\zeta.$$

Here, $W(h_1, \ldots, h_k)$ is the Wronskian of $h_1, \ldots, h_k$, and $W_m(h_1, \ldots, h_k)$ is the Wronskian of $h_1, \ldots, h_{m-1}, h_{m+1}, \ldots, h_k$. Moreover, given $\varepsilon > 0$, there exists $\delta_0 > 0$, $\delta_0 < \delta$, $n_0 \in \mathbb{N}$, and $M < \infty$ such that

$$\left| \frac{W_m(\zeta)}{W(\zeta)} \right| < M, \qquad |g_n(\zeta)| < \varepsilon, \ n > n_0, \ m = 1, \ldots, k, \ |\zeta| < \delta_0,$$

since g_n converges normally to g_0 and $g_0(0) = 0$. Therefore,

$$|\beta_{n,m}(z)| < M \cdot \varepsilon = \varepsilon^*, \qquad |z| < \delta_0, \ n > n_0.$$

Denote by $\mathcal{F}^*$ the class of associated combinations $\left\{ \sum_{m=1}^{k} \alpha_{n,m} h_m \right\}$ as determined by (4.29). Then for each $F_n \in \mathcal{F}^*$, there is a corresponding $f_n \in \mathcal{F}$ satisfying

$$
\begin{aligned}
|F_n(z) - f_n(z)| &\leq \sum_{m=1}^{k} |\beta_{n,m}(z) h_m(z)| \\
&< \varepsilon^* \cdot k \cdot \max_{1 \leq m \leq k} \left\{ \sup_{|z| \leq \delta_0} |h_m(z)| \right\} \\
&< \varepsilon_1, \qquad |z| < \delta_1(\varepsilon_1) < \delta_0, \ n > n_1(\varepsilon_1) > n_0.
\end{aligned}
$$

The normality of $\mathcal{F}$ will follow by demonstrating the normality of $\mathcal{F}^*$. To this end we show that the following auxiliary result holds:

Lemma 4.4.15 *Let $h_1(z), \ldots, h_k(z)$ be fixed linearly independent functions analytic in U. Let $\mathcal{F}^*$ be a family of analytic functions in U such that if $F \in \mathcal{F}^*$, then*

$$F(z) = \sum_{i=1}^{k} \alpha_i h_i(z)$$

for suitable constants $\alpha_i = \alpha_i(F)$. Moreover, suppose that for each $F \in \mathcal{F}^*$ there corresponds an analytic function $g = g(F)$ on U, such that $\{g\}$ is locally bounded, and

$$F(z) + g(z) = 0$$

has no solutions in U. Then $\mathcal{F}^*$ is normal in U.

Proof. Given $F \in \mathcal{F}^*$, $F(z) = \sum_{i=1}^{k} \alpha_i h_i(z)$, let $j = j(F)$ be the least integer such that $|\alpha_j| \geq |\alpha_i|$, $i \neq j$. Then write

$$F(z) = \alpha_j \sum_{i=1}^{k} \frac{\alpha_i}{\alpha_j} h_i(z), \qquad j = j(F).$$

From this formulation we see that $\mathcal{F}^*$ will be a locally bounded family and hence normal unless $\alpha_j \to \infty$ for some countably infinite subsequence

$\mathcal{F}_1^*$ of $\mathcal{F}^*$ and some fixed $j \in \{1, \ldots, K\}$. We may assume without loss of generality that $j = 1$. If $F \in \mathcal{F}_1^*$, then

$$F = \alpha_1 h_1 + \sum_{i=2}^{k} \frac{\alpha_i}{\alpha_1} h_i = \alpha_1 \left(h_1 + \sum_{i=2}^{k} \beta_i h_i \right), \qquad (4.30)$$

where $|\beta_i| \leq 1$, $i = 2, \ldots, k$. Let K be an arbitrary compact subset of U. Because the β_i corresponding to each $F \in \mathcal{F}_1^*$, are confined to a compact subset of $\mathbb{C}^{k-1}$, there is a convergent subsequence for $i = 2, \ldots, k$. Denote by

$$F^* = h_1 + \sum_{i=2}^{k} \beta_i^* h_i,$$

the limit of the corresponding subsequence of $\mathcal{F}_1^*$ under normal convergence in U. Then F^* is not identically zero since $\{h_1, \ldots, h_k\}$ are linearly independent, and so F^* does not identically vanish on K^0. Note that (4.30) may be written as

$$\frac{F}{\alpha_1} = h_1 + \sum_{i=2}^{k} \beta_i h_i.$$

Now, $F = F + g - g$ with $F + g$ nonvanishing on U. Moreover, as $|g(z)| < M$ and $\alpha_1 \to \infty$, then g/α_1 tends normally to zero and hence $(F + g)/\alpha_1$ tends normally to F^*. Since $F + g$ is nonzero and $F^* \not\equiv 0$, we have that F^* is nonzero on K^0 by the Hurwitz Theorem (§1.4).

Therefore, for K_1 a compact subset of K^0, there exists $\eta > 0$ with

$$\left| h_1 + \sum_{i=2}^{k} \beta_i h_i \right| > \eta, \qquad z \in K_1,$$

for an infinite subsequence of $\mathcal{F}_1^*$. Since $\alpha_1 \to \infty$, it follows from (4.30) that ∞ is a limit function of $\mathcal{F}^*$, and $\mathcal{F}^*$ is normal. This completes the proof of the theorem.

An Extension. Before proceeding with a full generalization of Theorem 4.4.14, we require an intermediate result which was not in the original 1969 paper of Drasin, but has been subsequently communicated to the author by him.

Theorem 4.4.16 *Let $\mathcal{F}$ be a family of zero-free analytic functions in U, and let $a_0(z), \ldots, a_{k-1}(z)$ be fixed analytic functions. Let $\mathcal{G}$ be the family of functions given by*

$$g(z) = f^{(k)}(z) + a_{k-1}(z) f^{(k-1)}(z) + \ldots + a_0(z) f(z), \qquad f \in \mathcal{F},$$

and assume that the zeros of $g(z) - 1$ are all of multiplicity $\geq p$, where $\frac{k+1}{p} = \tau < 1$. Then $\mathcal{F}$ is normal in U.

Proof. We first require the fact that:

> *If $\mathcal{G}$ is normal at the origin, then $\mathcal{F}$ is normal at the origin.*

As this follows similarly as in the same statement of Theorem 4.4.14, its validity may be assumed.

The proof of the theorem is broadly that of Theorem 4.4.14, subject to few modifications which we now indicate.

Proceeding as in the proof of Theorem 4.4.14, change (4.20) to

$$\left| \frac{g(\alpha) - 1}{g'(\alpha)} \right|^{1 - \frac{1}{p}} < 1. \tag{4.20'}$$

Next, write (4.26) as

$$
\begin{aligned}
T\left(r, \frac{1}{f_\alpha}\right) \;\leq\; & N\left(r, \frac{1}{f_\alpha}\right) + \bar{N}\left(r, \frac{1}{g_\alpha - 1}\right) \\
& + A \log^+ T(R, f_\alpha) + B \log \frac{1}{R - r} \\
& + \log \left| \frac{g(\alpha) - 1}{g'(\alpha)} \right| + C, \qquad \tfrac{\delta}{2} < r < R < \delta.
\end{aligned} \tag{4.26'}
$$

Observe that

$$
\begin{aligned}
\bar{N}\left(r, \frac{1}{g_\alpha - 1}\right) \;\leq\; & \frac{1}{p} N\left(r, \frac{1}{g_\alpha - 1}\right) \leq \frac{1}{p} T\left(r, \frac{1}{g_\alpha - 1}\right) \\
= \; & \frac{1}{p}\left(T(r, g_\alpha - 1) + \log \frac{1}{|g(\alpha) - 1|} \right) \\
\leq \; & \frac{1}{p}\left(T(r, g_\alpha) + C + \log \frac{1}{|g(\alpha) - 1|} \right).
\end{aligned}
$$

In order to deal with the term $T(r, g_\alpha)$, we have by virtue of the proof of Corollary 4.4.13 that

$$T(r, g_\alpha) \leq (k + 1)T(r, f_\alpha) + A \log^+ T(R, f_\alpha) + B \log \frac{1}{R - r} + C,$$

for an infinite subfamily $\mathcal{F}_{5'} \subseteq \mathcal{F}_5$, $\mathcal{G}_{5'} \subseteq \mathcal{G}_5$, $\tfrac{\delta}{2} < r < R < \delta$.

From the two preceding inequalities, (4.26') becomes (since $\mathcal{F}$ is zero-free)

$$
\begin{aligned}
T\left(r, \frac{1}{f_\alpha}\right) \;\leq\; & \frac{k + 1}{p} T(r, f_\alpha) + A \log^+ T(R, f_\alpha) + B \log \frac{1}{R - r} \\
& + \log \left| \frac{g(\alpha) - 1}{g'(\alpha)} \right|^{1 - \frac{1}{p}} + C, \qquad \tfrac{\delta}{2} < r < R < \delta.
\end{aligned} \tag{4.27'}
$$

Finally, for (4.28) write

$$T(r, f_\alpha) \le (1 + \varepsilon) T\left(r, \frac{1}{f_\alpha}\right),\qquad (4.28')$$

for $f \in \mathcal{F}_{6'} \subseteq \mathcal{F}_{5'}$, $\frac{\delta}{2} < r < R < \delta$. Choosing ε sufficiently small such that $\left(\frac{k+1}{p}\right)(1 + \varepsilon) < 1$, and combining (4.28') with (4.27'), the proof is completed as in Theorem 4.4.14.

The generalization of Theorem 4.14 we now seek to prove is the following.

Theorem 4.4.17 *Let $\mathcal{F}$ be a family of analytic functions in U. If the zeros of each $f \in \mathcal{F}$ are of multiplicity $\ge m$, and the zeros of $g(z) - 1$ are of multiplicity $\ge p$, where*

$$\frac{k+1}{m} + \frac{k+1}{p} = \tau < 1,$$

$g(z) = f^{(k)}(z) + a_{k-1}(z)f^{(k-1)}(z) + \ldots + a_0(z)f(z)$, *with $a_0(z), \ldots, a_{k-1}(z)$ analytic in U, then $\mathcal{F}$ is normal.*

Remark. If j of the functions $a_0(z), \ldots, a_{k-1}(z)$ are identically zero, then it suffices to assume

$$\frac{k+1}{m} + \frac{k-j+1}{p} < 1.$$

The case $j = k$ was derived by Yang and Chang [1965] along the lines of the proof of Theorem 4.5.5. An application of this particular case is the proof of Theorem 2.10.3.

Digressing for a moment, we note that this theorem permits a further illustration of the Bloch Principle. Indeed, a well-known result in function theory due to Hayman [1959] ($n \ge 2$) and Clunie [1967] ($n = 1$) states that: *If $f(z)$ is an entire function satisfying*

$$f'(z)f(z)^n \ne 1,\qquad (4.31)$$

for all $z \in \mathbb{C}$, then $f \equiv$ constant.

There is a normal family analogue, namely,

Theorem 4.4.18 *Let $\mathcal{F}$ be a family of analytic functions in U satisfying*

$$h'(z)h(z)^n \ne 1,\qquad z \in U,$$

where $n \ge 1$ is a fixed integer. Then $\mathcal{F}$ is normal.

Proof. For $n \ge 2$, the result is a direct consequence of the preceding theorem as well as the Yang and Chang version. To see this, let

$$\mathcal{F}^* = \left\{ f(z) = \left(\frac{1}{n+1}\right) h(z)^{n+1} : h \in \mathcal{F} \right\},$$

and let $g(z) = f'(z)$. Then $k = 1$, and the zeros of $f \in \mathcal{F}^*$ are of multiplicity $\geq n+1 = m$. As $g(z) - 1$ has no zeros in U, and $\frac{k+1}{m} \leq \frac{2}{3} < 1$, the theorem implies $\mathcal{F}^*$ is normal, and so is $\mathcal{F}$.

The case $n = 1$ has been considered by Oshkin [1982] and is a consequence of a generalization due to Schwick [1989] (cf. Theorem 4.5.3). Pang [1989] has also given a proof via the Heuristic Principle (cf. Note subsequent to Theorem 4.5.2). An extension by Hua and Chen [preprint] shows that: $\mathcal{F}$ *is normal in* U *if for some constant* $M > 0$, $|h'(z)| \leq M$ *on* $\{z \in U : h'(z)h(z)^n = 1\}$, $n \geq 1$.

Proof of Theorem 4.4.17   Once more we require:

If $\mathcal{G}$ *is normal at the origin, then* $\mathcal{F}$ *is normal at the origin.*

The proof of this assertion is reminiscent of that required in Theorem 4.4.14. Assume $g_n \to g_0$ normally in a neighbourhood of the origin.

Case 1. $g_0(0) \neq 0$. As a zero of the corresponding f_n is also a zero of g_n (since $m > k + 1$), the origin cannot be a limit point of zeros of the f_n's by the Hurwitz Theorem. Then the argument of Case 1 of Theorem 4.4.14 may be applied directly.

Case 2. $g_0(0) = 0$, but for some $\delta > 0$

$$n\left(r, \frac{1}{f_n}\right) = 0, \qquad r < \delta,$$

for infinitely many f_n. Then the argument of Case 2 of Theorem 4.4.14 is applicable.

Case 3. $g_0(0) = 0$ and there is a sequence $z_n \to 0$ with $f_n(z_n) = 0$, and hence $g_n(z_n) = 0$.

Again, denoting by $h_1(z), \ldots, h_k(z)$, the linearly independent solutions of the differential equation $g(z) = 0$, we obtain as before

$$f_n(z) = f_{n,h} + \sum_{m=1}^{k} \left(\int_{z_n}^{z} \frac{W_m(\zeta)}{W(\zeta)} g_n(\zeta)\, d\zeta \right) h_m(z), \qquad (4.32)$$

where $f_{n,h}(z) = \sum_{m=1}^{k} \alpha_{n,m} h_m(z)$ is a solution of the homogeneous equation $g(z) = 0$. Let us assume for the moment that

$$f_{n,h} \equiv 0, \qquad n > n_0, \qquad (4.33)$$

so that (4.32) becomes

$$f_n(z) = \sum_{m=1}^{k} \left(\int_{z_n}^{z} \frac{W_m(\zeta)}{W(\zeta)} g_n(\zeta)\, d\zeta \right) h_m(z). \qquad (4.34)$$

For $\delta > 0$, one can find $M < \infty$ such that

$$|g_n(z)| < M, \quad \left|\frac{W_m(z)}{W(z)}\right| < M, \quad |h_m(z)| < M,$$

$|z| < \delta$, $m = 1, \ldots, k$, $n > n_0$. Then

$$|f_n(z)| < M^*,$$

$n = 1, 2, 3, \ldots$, and we conclude that $\{f_n\}$ is normal at the origin.

In order to show (4.33), take $f_n = f$, $f_{n,h} = f_h$, $g_n = g$ and rewrite (4.32) as

$$f(z) = f_h(z) + \sum_{m=1}^{k} \beta_m(z) h_m(z). \qquad (4.35)$$

Clearly, $\beta_m(z_n) = 0$, $m = 1, \ldots, k$. Differentiating equation (4.35) leads to

$$f(z) = f_h'(z) + \sum_{m=1}^{k} [\beta_m(z) h_m'(z) + \beta_m'(z) h_m(z)]. \qquad (4.36)$$

From the method of variation of parameters, the $\beta_m(z)$ are subject to the following constraints:

$$\beta_1'(z) h_1(z) \quad + \quad \beta_2'(z) h_2(z) \quad + \quad \cdots \quad + \quad \beta_k'(z) h_k(z) = 0,$$
$$\vdots$$
$$\beta_1'(z) h_1^{(k-2)}(z) \quad + \quad \beta_2'(z) h_2^{(k-2)}(z) \quad + \quad \cdots \quad + \quad \beta_k'(z) h_k^{(k-2)}(z) = 0,$$
$$\beta_1'(z) h_1^{(k-1)}(z) \quad + \quad \beta_2'(z) h_2^{(k-1)}(z) \quad + \quad \cdots \quad + \quad \beta_k'(z) h_k^{(k-1)}(z) = g(z).$$

By virtue of the first constraint, (4.36) reduces to

$$f'(z) = f_h'(z) + \sum_{m=1}^{k} \beta_m(z) h_m'(z). \qquad (4.37)$$

Since z_n has multiplicity $\geq m > k + 1$ and $\beta_m(z_n) = 0$, we obtain from (4.37)

$$0 = f'(z_n) = f_h'(z_n).$$

Considering successive derivatives together with successive constraints coupled with the fact that $\beta_m(z_n) = 0$ and $f^{(p)}(z_n) = 0$, $p = 1, \ldots, k - 1$, we further deduce that

$$0 = f_h''(z_n) = \ldots = f_h^{(k-1)}(z_n).$$

Since $f_h(z_n) = 0$ as well, the conclusion (4.33) follows from the uniqueness of the solution to the homogeneous equation $g(z) = 0$ in a neighbourhood of z_n. The normality of $\mathcal{F}$ can thus be derived from that of $\mathcal{G}$.

On the other hand, let $\mathcal{F}_1$ be an arbitrary sequence in $\mathcal{F}$; it will be shown that $\mathcal{F}_1$ is normal in a neighbourhood of $z_0 = 0$. Unless z_0 is a limit point of zeros of $\mathcal{F}_1$, the theorem is a consequence of Theorem 4.4.16.

Now, if $f(\alpha) = f_\alpha(0) \neq 0$, we have

$$N\left(r, \frac{1}{f_\alpha}\right) - N^*\left(r, \frac{1}{g_{\alpha'}}\right) = (k+1)\bar{N}\left(r, \frac{1}{f_\alpha}\right) \leq \frac{k+1}{m} N\left(r, \frac{1}{f_\alpha}\right)$$

$$\leq \frac{k+1}{m} T\left(r, \frac{1}{f_\alpha}\right) = \frac{k+1}{m}\left(T(r, f_\alpha) - \log|f(\alpha)|\right), \quad (4.38)$$

where $N^*\left(r, \frac{1}{g_{\alpha'}}\right)$ counts the zeros which are common to both g_α' and the corresponding f_α. Moreover,

$$N\left(r, \frac{1}{g_\alpha - 1}\right) - N^{**}\left(r, \frac{1}{g_{\alpha'}}\right) \leq \frac{1}{p} N\left(r, \frac{1}{g_\alpha - 1}\right)$$

$$\leq \frac{1}{p}\left(T(r, g_\alpha) - \log|g(\alpha) - 1| + D\right), \quad (4.39)$$

where $N^{**}\left(r, \frac{1}{g_{\alpha'}}\right)$ counts the zeros which are common to both g_α' and $g_\alpha - 1$.

Some of the analysis carried out in the proof of Theorem 4.4.14 may be revived at this point as we are only considering a neighbourhood of the origin, $|z| < \delta$, and $|\alpha| \leq r_0 < \delta$. Disregarding any consideration of subfamilies for the moment, we formally put (4.38) and (4.39) into (4.26), and invoking $T(r, f_\alpha) = T\left(r, \frac{1}{f_\alpha}\right) + \log|f(\alpha)|$ gives

$$\begin{aligned} T(r, f_\alpha) \quad < \quad & \frac{k+1}{m}\left(T(r, f_\alpha) - \log|f(\alpha)|\right) + \log|f(\alpha)| \\ & + \frac{1}{p}\left(T(r, g_\alpha) - \log|g(\alpha) - 1| + D\right) \\ & + A\log^+ T(R, f_\alpha) + B\log\frac{1}{R-r} + \log\left|\frac{g(\alpha) - 1}{g'(\alpha)}\right| + C, \\ & \hspace{5cm} \frac{\delta}{2} < r < R < \delta. \quad (4.40) \end{aligned}$$

Applying again the analysis of Corollary 4.4.13 yields

$$T(r, g_\alpha) < (k+1)\, T(r, f_\alpha) + A\log^+ T(r, f_\alpha) + B\log\frac{1}{R-r} + C,$$

$\frac{\delta}{2} < r < R < \delta$. Putting this estimate of $T(r, g_\alpha)$ into (4.40) gives, upon simplifying,

$$\begin{aligned} (1-\tau)\, T(r, f_\alpha) \quad < \quad & A\log^+ T(R, f_\alpha) + B\log\frac{1}{R-r} + C \\ & + \log\left|[f(\alpha)]^{1-\frac{(k+1)}{m}} \cdot [g(\alpha) - 1]^{\frac{p-1}{p}} \cdot [g'(\alpha)]^{-1}\right|, \\ & \hspace{4cm} \frac{\delta}{2} < r < R < \delta, \quad (4.41) \end{aligned}$$

for f belonging to an infinite subset $\mathcal{F}_2 \subseteq \mathcal{F}_1$. Here the constants only depend on δ and the family $\mathcal{F}_1$. Observe that $\mathcal{F}$ will be normal in a neighbourhood of the origin by applying Lemma 4.4.4 to (4.41) if there exists a sequence $\{\alpha_n\}$ such that $\alpha_n \to 0$ and

$$\left| f(\alpha)^{\frac{m-k-1}{m}} [g(\alpha) - 1]^{\frac{p-1}{p}} \cdot g'(\alpha)^{-1} \right| < M, \tag{4.42}$$

$\alpha = \alpha_n$, $f = f_n \in \mathcal{F}_2$ corresponding to $g = g_n \in \mathcal{G}_2$.

To this end, suppose that f_n has a zero z_n, with $z_n \to 0$. Since $m > k+1$, z_n is a zero of both g_n and g'_n. For each z_n, let ζ_n be determined by minimizing $|z_n - \zeta_n|$, subject to the condition $|g'_n(\zeta_n)| = 1$; if no such ζ_n exists for infinitely many n, then the corresponding g'_n would be uniformly bounded, and hence these g_n would form a normal family by Theorem 2.2.6 (or Marty's theorem), implying the normality of $\mathcal{F}$.

Therefore, consider $|z_n - \zeta_n| = \delta_n$, so that if $|z - z_n| < \delta_n$,

$$|g_n(z)| \le \int_{z_n}^{z} |g'_n(\xi)||d\xi| < \delta_n. \tag{4.43}$$

Suppose that a subsequence $\{\delta_{n_k}\}$ satisfies $\delta_{n_k} > \eta > 0$. Then from (4.43), the definition of δ_{n_k}, and the fact that $z_{n_k} \to 0$, we get

$$|g_{n_k}(z)| \le 1, \qquad |z| < \frac{\eta}{2}.$$

Thus $\mathcal{G}$, and whence $\mathcal{F}$, is normal in a neighbourhood of the origin.

Therefore, assume that $\delta_n \to 0$. It follows from (4.34) and (4.43) that $f_n(\zeta_n) \to 0$. Furthermore,

$$\left| \frac{g_n(\zeta_n)}{g'_n(\zeta_n)} \right| \le \int_{z_n}^{\zeta_n} |g'_n(\xi)||d\xi| \le |z_n - \zeta_n| = \delta_n.$$

We deduce that the expression on the left of (4.42) actually tends to zero as $n \to \infty$, for $\alpha = \zeta_n$. The proof of the theorem is herewith complete.

A New Normality Criterion.

Another result due to Hayman [1959] is the following:

If $f(z)$ is entire (meromorphic in $\mathbb{C}$), and $\nu \ge 3$ ($\nu \ge 5$) is a fixed integer, $a \ne 0$, and

$$f'(z) - af(z)^{\nu} \ne b$$

for some $b \in \mathbb{C}$, then $f \equiv$ constant.

Mues [1979] has given counterexamples to show that if $f(z)$ is meromorphic in $\mathbb{C}$ and $\nu = 3, 4$, the result is no longer valid. In terms of normal families we have:

Theorem 4.4.19 *Let $\mathcal{F}$ be a family of analytic (meromorphic) functions in U such that for fixed $\nu \geq 3$ ($\nu \geq 5$) and $a \neq 0$*

$$f'(z) - af(z)^\nu \neq b, \qquad f \in \mathcal{F},$$

for some $b \in \mathbb{C}$. Then $\mathcal{F}$ is normal.

The analytic case is due to Drasin [1969] (cf. also Pang [1990] for $\nu = 3$, Ye [1991] for $\nu = 2$, Chen and Hua [preprint] for a generalization with a and b replaced by meromorphic functions). The meromorphic case is due to Langley [1984], S. Li [1984], and X. Li [1985]. Moreover, despite the counterexamples of Mues, Pang [1990] has demonstrated the validity of Theorem 4.4.19 for a family $\mathcal{F}$ of meromorphic functions and $\nu \geq 4$. This was achieved via the Zalcman Lemma with a modification of (4.1). With the additional assumption that the zeros of every $f \in \mathcal{F}$ have multiplicity not less than 2, Pang proved the case $\nu = 3$. However, this extra assumption on the zeros of $f \in \mathcal{F}$ is not necessary if a conjecture made by Hayman is true (cf. Note after Theorem 4.5.2). For a related result with $\nu = 3$, and an assumption on the degree of the poles of each function f, see Chen and Gu [preprint].

We present the Drasin proof of the analytic case as we now have all the necessary machinery at our disposal.

To this end, define the family

$$\mathcal{H} = \left\{ h(z) = \frac{af(z)^\nu}{f'(z) - b - af(z)^\nu} : f \in \mathcal{F} \right\}.$$

Then $h \in \mathcal{H}$ is analytic by the assumptions of the theorem. First it will be shown that:

If $\mathcal{H}$ is normal, then $\mathcal{F}$ is normal.

Proof. Assuming $h(\alpha) \neq -1$, routine calculations give

$$\nu T(r, f_\alpha) = T(r, f_\alpha^\nu) = T\left(r, \frac{f'_\alpha - b}{a} \cdot \frac{h_\alpha}{h_\alpha + 1} \right)$$

$$\leq T(r, f_\alpha) + m\left(r, \frac{f'_\alpha}{f_\alpha} \right) + T(r, h_\alpha) - \log|h(\alpha) + 1| + K(a, b),$$

so that

$$T(r, f_\alpha)$$
$$\leq \frac{1}{\nu - 1} \left\{ m\left(r, \frac{f'_\alpha}{f_\alpha} \right) + T(r, h_\alpha) - \log|h(\alpha) + 1| + K(a, b) \right\}. \quad (4.44)$$

Let $\mathcal{F}_1$ be an infinite sequence of elements of $\mathcal{F}$, and $\mathcal{H}_1$ the corresponding sequence in $\mathcal{H}$. Letting h^* denote a limit function of $\mathcal{H}_1$, there are three cases to consider.

Case 1. $h^* \not\equiv -1, \infty$. Suppose that $\{h_n\} = \mathcal{H}_2 \subseteq \mathcal{H}_1$ converges to h^*. By our assumption, there is some $M < \infty$, $r_0 < 1$, $\alpha_n = \alpha(h_n)$ with $|\alpha| \leq r_0$ such that

$$-\log|h_n(\alpha_n) + 1| < M,$$

for $h_n \in \mathcal{H}_3 \subseteq \mathcal{H}_2$. Furthermore, unless $\mathcal{F}_3$ is normal, the estimate of Lemma 4.4.8 is applicable to the logarithmic derivative term in (4.44) to give

$$T(r, f_{n,\alpha_n})$$
$$\leq \frac{1}{\nu - 1} \left\{ A \log^+ T(R, f_{n,\alpha_n}) + B \log \frac{1}{R - r} + T(r, h_{n,\alpha_n}) + M \right\},$$
$$\tfrac{1}{2} < r < R < 1, \quad (4.45)$$

for $\mathcal{F}_4 \subseteq \mathcal{F}_3$.

At this stage, apply the estimate of Lemma 4.4.4 to (4.45) with $V(r) = T(r, f_{n,\alpha_n})$, $\gamma(r) = \frac{1}{\nu-1}T(r, h_{n,\alpha_n})$, increasing B if necessary. Then

$$T(r, f_{n,\alpha_n}) < A + B \cdot T(R, h_{n,\alpha_n}) + C \log \frac{1}{R - r}, \quad \tfrac{1}{2} < r < R < 1, \quad (4.46)$$

$f_n \in \mathcal{F}_4$, $h_n \in \mathcal{H}_4$.

Note that for any $\rho < 1$,

$$T(\rho, h_{n,\alpha_n}) \to T(\rho, h_\alpha^*) \qquad \text{as } n \to \infty.$$

This fact, together with (4.46) and Lemma 4.4.1, implies the normality of $\mathcal{F}_4$ and hence of $\mathcal{F}$.

Case 2. $h^* \equiv -1$. Given any $r_0 < 1$ and $\varepsilon > 0$, then

$$\left| \frac{f'(z) - b}{af(z)^\nu} \right| < \varepsilon, \qquad |z| < r_0,$$

holds for f belonging to an infinite subfamily $\mathcal{F}_0 \subseteq \mathcal{F}_1$.

Let us show that the following argument holds: *For each r_0, there exists $\eta = \eta(r_0)$, such that if $|z_1| < r_0$, $|z_2| < r_0$, with*

$$|f(z_1)| \leq 1 \tag{4.47}$$

and

$$|f(z_2)| \geq 2 \tag{4.48}$$

for $f \in \mathcal{F}_0$, then

$$|z_1 - z_2| \geq \eta. \tag{4.49}$$

In fact, choose z_1 and z_2 in accordance with (4.47) and (4.48), and we suppose that $|z_1 - z_2|$ is minimized subject to these conditions. Join z_1 and z_2 with a line segment γ. Since γ lies in $|z| < r_0$, we have for z on γ,

$$|f'(z)| \leq \varepsilon|af(z)^\nu| + |b| \leq K\varepsilon + \beta, \qquad K = K(a, \nu), \quad \beta = |b|,$$

in view of $|f(z)| \le 2$. Therefore

$$1 \le |f(z_2) - f(z_1)| \le \int_\gamma |f'(z)||dz| \le (K\varepsilon + \beta)|z_1 - z_2|,$$

proving (4.49).

This then means that in a prescribed neighbourhood of each z in $|z| < r_0$, every $f \in \mathcal{F}_0$ is either uniformly bounded above or below, so that $\mathcal{F}$ is normal in U.

Case 3. $h^* \equiv \infty$. Then for any $r_0 < 1$, $\varepsilon > 0$,

$$\left|\frac{f'(z) - b}{af(z)^\nu}\right| < 1 + \varepsilon, \qquad |z| < r_0,$$

for infinitely many $f \in \mathcal{F}$. Applying the argument of Case 2 again gives the normality of $\mathcal{F}$.

Proof of Theorem 4.4.19 Let $\mathcal{F}_1$ be a sequence in $\mathcal{F}$, and $\mathcal{H}_1$ the corresponding sequence in $\mathcal{H}$. We show that $\mathcal{H}_1$ is normal in a neighbourhood of the origin. Initially assume $|\alpha| < \frac{1}{4}$ and $|\beta|$ small; they will be made more precise in the sequel.

Let us suppose that $\mathcal{H}_1$ is not normal at the origin. Then applying the inequality (4.6) to the functions h_α and invoking Lemma 4.4.9 to estimate the logarithmic derivative terms yields an infinite subfamily $\mathcal{H}_2 \subseteq \mathcal{H}_1$ whose members satisfy

$$m(r, h_\alpha) + m\left(r, \frac{1}{h_\alpha}\right) + m\left(r, \frac{1}{h_\alpha + 1}\right) \le 2T(r, h_\alpha) - N\left(r, \frac{1}{h_{\alpha'}}\right)$$

$$+ A\log^+ T(R, h_\alpha) + B\log\frac{1}{R - r} + \log\left|\frac{1}{h'(\alpha)}\right| + C,$$
$$0 < \delta_1 < r < R < 1. \quad (4.50)$$

It is tacitly assumed that $h(\alpha) \ne 0, -1$, $h'(\alpha) \ne 0$.

According to Remark 4.4.7 and Lemma 4.4.6, given $\varepsilon > 0$ there exists $\delta_2 < \delta_1/2$ such that for $|\beta| < \delta_2$ and r as in (4.50)

$$(1 - \varepsilon)\, m\left(r, \frac{1}{h_\alpha + 1} \circ \psi_\beta\right) \le m\left(r, \frac{1}{h_\alpha + 1}\right). \quad (4.51)$$

Furthermore,

$$(1 - \varepsilon)\, T(r, h_\alpha) \le T(r, h_\alpha \circ \psi_\beta), \qquad |\beta| < \delta_2,\ \delta_1 < r, \quad (4.52)$$

follows likewise as h is analytic, with ε yet to be determined.

Adding the terms

$$N(r, h_\alpha) + (1 - \varepsilon)N\left(r, \frac{1}{h_\alpha + 1} \circ \psi_\beta\right)$$

to both sides of (4.50) and using (4.51) and (4.52), along with the First Fundamental Theorem leads to the inequality

$$(1-\varepsilon)^2\, T(r,h_\alpha) \le \bar{N}\!\left(r,\frac{1}{h_\alpha}\right) + (1-\varepsilon)N\!\left(r,\frac{1}{h_\alpha+1}\circ\psi_\beta\right)$$

$$+A\log^+ T(R,h_\alpha) + B\log\frac{1}{R-r} + (1-\varepsilon)\log|(h_\alpha+1)(-\beta)|$$

$$+\log\left|\frac{h(\alpha)}{h'(\alpha)}\right| + C, \qquad |\beta|<\delta_2,\ \ \delta_1<r<R<1. \tag{4.53}$$

In order to treat the first term on the right side of (4.53), note that the zeros of h_α are all of multiplicity $\ge \nu$, so that

$$\bar{N}\!\left(r,\frac{1}{h_\alpha}\right) \le \frac{1}{\nu}N\!\left(r,\frac{1}{h_\alpha}\right) \le \frac{1}{\nu}\{T(r,h_\alpha) - \log|h(\alpha)|\}. \tag{4.54}$$

Considering the next term

$$N\!\left(r,\frac{1}{h_\alpha+1}\circ\psi_\beta\right) = N\!\left(r,\left\{\frac{af^\nu}{h_\alpha}\cdot\frac{1}{f'_\alpha-b}\right\}\circ\psi_\beta\right) \le N\!\left(r,\frac{1}{f'_\alpha-b}\circ\psi_\beta\right)$$

$$\le T(r,(f'_\alpha-b)\circ\psi_\beta) - \log|(f'_\alpha-b)(-\beta)|$$

$$\le T(r,f_\alpha\circ\psi_\beta) + m\!\left(r,\frac{f'_\alpha}{f_\alpha}\circ\psi_\beta\right)$$

$$-\log|(f'_\alpha-b)(-\beta)| + C. \tag{4.55}$$

As in (4.51) and (4.52),

$$T(r,f_\alpha\circ\psi_\beta) < (1+\varepsilon)\,T(r,f_\alpha), \tag{4.56}$$

$$m\!\left(r,\frac{f'_\alpha}{f_\alpha}\circ\psi_\beta\right) < (1+\varepsilon)\,m\!\left(r,\frac{f'_\alpha}{f_\alpha}\right). \tag{4.57}$$

Unless $\mathcal{F}_2$ is normal in a neighbourhood of the origin, we apply Lemma 4.4.9 to estimate the logarithmic derivative term in (4.57). Together with the estimates (4.54) – (4.57) in (4.53), we deduce that for an infinite sub-family $\mathcal{F}_3 \subseteq \mathcal{F}_2$ ($\mathcal{H}_3 \subseteq \mathcal{H}_2$),

$$\left((1-\varepsilon)^2 - \frac{1}{\nu}\right)T(r,h_\alpha)$$

$$\le (1+\varepsilon)\left\{T(r,f_\alpha) + A\log^+ T(\rho,f_\alpha) + B\log\frac{1}{\rho-r}\right\} + A\log^+ T(r,h_\alpha)$$

$$+B\log\frac{1}{R-r} + (1-\varepsilon)\log\left|\frac{(h_\alpha+1)(-\beta)}{(f'_\alpha-b)(-\beta)}\right| + \log\left|\frac{h(\alpha)^{1-\frac{1}{\nu}}}{h'(\alpha)}\right| + C,$$

$$\delta_1<r<\rho,\ \ r<R<1,\ \ |\beta|<\delta_2. \tag{4.58}$$

Assuming $\mathcal{F}_3$ is not normal at the origin, and hence likewise for $\mathcal{H}_3$, by Marty's theorem there exists a sequence of functions $\mathcal{H}_4 = \{h_n\} \subseteq \mathcal{H}_3$

and points $\alpha_n = \alpha(h_n)$, $\alpha_n \to 0$, such that (with $h = h_n$, $\alpha = \alpha_n$, $n = 1, 2, 3, \ldots$)

$$\left| \frac{h(\alpha)^{1-\frac{1}{\nu}}}{h'(\alpha)} \right| < 1, \tag{4.59}$$

where $h(\alpha) \neq 0, -1$, $h'(\alpha) \neq 0$.

To deal with the preceding term in (4.58), suppose that $\{f' - b - af^\nu\}$ tended normally to zero in a neighbourhood of the origin, for $f \in \mathcal{F}_4$. Then a repetition of the argument that (4.49) follows from (4.47) and (4.48) would show that $\mathcal{F}_4$ would be normal at the origin. Hence we may presume the supposition to be false, and consequently there exists a subsequence $\{f_n\} = \mathcal{F}_5 \subseteq \mathcal{F}_4$ ($\mathcal{H}_5 \subseteq \mathcal{H}_4$) and a sequence of points $\{\beta_n\}$ with $\beta_n \to 0$ corresponding to f_{n,α_n} satisfying

$$\left| \frac{h_{n,\alpha_n} + 1}{f'_{n,\alpha_n} - b}(-\beta_n) \right| = \left| \frac{1}{f'_{n,\alpha_n} - b - af^\nu_{n,\alpha_n}}(-\beta_n) \right| < M,$$
$$n = 1, 2, 3, \ldots . \tag{4.60}$$

In view of (4.59) and (4.60), the inequality (4.58) may be simplified, by taking $f \in \mathcal{F}_5$ with the corresponding $h \in \mathcal{H}_5$, to

$$\left((1 - \varepsilon)^2 - \frac{1}{\nu} \right) T(r, h_\alpha)$$

$$\leq (1 + \varepsilon) \left\{ T(r, f_\alpha) + A \log^+ T(\rho, f_\alpha) + B \log \frac{1}{\rho - r} \right\}$$

$$+ A \log^+ T(R, h_\alpha) + B \log \frac{1}{R - r} + C,$$
$$\delta_1 < r < \rho, \ r < R < 1. \tag{4.61}$$

Next, we wish to majorize $T(r, f_\alpha)$ in terms of $T(r, h_\alpha)$ by replacing f_α by $f_\alpha \circ \psi_\beta$ in (4.44). We may assume that the functions in $\mathcal{H}_5$ do not tend normally to -1 or ∞ in a neighbourhood of the origin, for otherwise $\mathcal{F}_5$ would be normal there. Then one can find a sequence of points $\{\beta_n\}$ with $\beta_n \to 0$ associated with h_{n,α_n}, where $\{h_n\} = \mathcal{H}_6 \subseteq \mathcal{H}_5$ and

$$-\log |h_{n,\alpha_n}(-\beta_n) + 1| < M.$$

Hence (4.44) becomes, on noting (4.56) and (4.57),

$$T(r, f_\alpha) \leq \frac{(1 + \varepsilon)^2}{\nu - 1} \left\{ m\left(r, \frac{f'_\alpha}{f_\alpha} \right) + T(r, h_\alpha) + M \right\},$$
$$\delta_1 < r < 1, \ f \in \mathcal{F}_6. \tag{4.62}$$

Finally, unless $\mathcal{F}_6$ is normal at the origin,

$$m\left(r, \frac{f'_\alpha}{f_\alpha} \right) < A \log^+ T(R, f_\alpha) + B \log \frac{1}{R - r} + C, \quad \delta_1 < r < R < 1, \tag{4.63}$$

for $f \in \mathcal{F}_7$, an infinite subsequence of $\mathcal{F}_6$. Then (4.62) simplifies by virtue of Lemma 4.4.4 and (4.63) with R replacing r:

$$T(R, f_\alpha) \le A \cdot T(R, h_\alpha) + B \log \frac{1}{\rho' - R} + C, \qquad \delta_1 < R < \rho' < 1. \quad (4.64)$$

Putting (4.64) into (4.63) and choosing $R = \frac{r+\rho'}{2}$ $\left(\text{so that } \frac{1}{\rho'-R} = \frac{1}{R-r}\right)$, (4.62) now reads, noting the relation amongst r, R, ρ',

$$T(r, f_\alpha) \le \frac{(1+\varepsilon)^2}{\nu - 1} T(r, h_\alpha) + A \log^+ T(R, h_\alpha) + B \log \frac{1}{R-r} + C,$$
$$\delta_1 < r < R < 1, \ f \in \mathcal{F}_7.$$

By inserting this last estimate into (4.61) taking $\rho = R$, coupled with (4.64),

$$\left((1-\varepsilon)^2 - \frac{1}{\nu} - \frac{(1+\varepsilon)^3}{\nu - 1}\right) T(r, h_\alpha)$$
$$\le A \log^+ T(R, h_\alpha) + B \log \frac{1}{R-r} + C, \quad \delta_1 < r < R < 1. \quad (4.65)$$

Choosing $\varepsilon > 0$ sufficiently small, the normality of $\mathcal{H}_7$ results from an application of Lemma 4.4.4, and the normality of $\mathcal{F}_7$ at the origin is achieved, completing the proof.

4.5   Further Results

Since the publication of Drasin's 1969 paper, there have been numerous extensions of the basic results to the meromorphic (and some analytic) cases. A considerable amount of progress has been made by the "Chinese School", principally, Yang, Chang, Ku, Chen, Li, Hua, and Pang, their work having its origins rooted in the works of Hiong and Chuang. Outside this sphere, besides the developments pertaining to the Robinson-Zalcman Heuristic Principle, is some very recent work of Schwick, amongst others, based somewhat on the Drasin approach coupled with the Zalcman Lemma.

As in the case for entire functions, a meromorphic function satisfying

$$f'(z)f(z)^n \ne 1 \qquad (n \ge 2)$$

in $\mathbb{C}$ is constant (Hayman [1959] for $n \ge 3$, Mues [1979] for $n = 2$, and conjectured true by Hayman [1967], p. 7, for $n \ge 1$). The meromorphic equivalent of Theorem 4.4.18 for $n \ge 3$ is a consequence of the following result of Ku [1978] pertaining to Theorem 4.4.17.

Theorem 4.5.1 *Let $\mathcal{F}$ be a family of meromorphic functions in Ω. Suppose that for each $f \in \mathcal{F}$, all the poles have multiplicity $\ge s$ (≥ 1), all the* *zeros have multiplicity* $\geq m$ (≥ 2), *and there are distinct nonzero numbers* $b_1, \ldots, b_q$ *such that all the zeros of* $f^{(k)}(z) - b_j$ *have multiplicity* $\geq n_j$ (≥ 2), $j = 1, \ldots, q$, *where*

$$\frac{kq+1}{m} + \sum_{j=1}^{q} \frac{1}{n_j} + \frac{1}{s}\left(1 + k\sum_{j=1}^{q}\frac{1}{n_j}\right) < q.$$

Then $\mathcal{F}$ *is normal in* Ω.

An immediate consequence is

Theorem 4.5.2 *If* $\mathcal{F}$ *is a family of meromorphic functions in* Ω *such that* $f'(z)f(z)^n \neq 1$ $(n \geq 3)$ *for all* $f \in \mathcal{F}$, *then* $\mathcal{F}$ *is normal.*

In fact, upon setting $q = 1$ in the preceding theorem, the result is easily deduced as in the analytic case (Theorem 4.4.18).

Theorem 4.5.2 has been improved by Pang [1989] to include the case $n \geq 2$. The proof is via the Heuristic Principle and the extended definition of a normal property with $k = \frac{1}{n+1}$ (n fixed). To this end, define $\mathcal{P}$ by $\langle f, \Omega \rangle \in \mathcal{P}$ if either $f'(z)f(z)^n \neq 1$, $z \in \Omega$, or $f \equiv \infty$. Verifying conditions (i) and (ii') of Definition 4.11 is straightforward enough. Now suppose $\langle f_j, \Omega_j \rangle \in \mathcal{P}$, for $\Omega_1 \subseteq \Omega_2 \subseteq \ldots$ and $\mathbb{C} = \cup_{j=1}^{\infty}\Omega_j$, such that $f_n \to f$ spherically uniformly on compact subsets of $\mathbb{C}$. We may assume f is nonconstant, for otherwise $\langle f, \mathbb{C} \rangle \in \mathcal{P}$ trivially. If $\langle f, \mathbb{C} \rangle \notin \mathcal{P}$, then there is a point $z_0 \in \mathbb{C}$ at which $f'(z_0)f(z_0)^n = 1$. Since f is bounded analytic in some closed neighbourhood $K(z_0; r)$, $f_j'f_j^n \to f'f^n$ uniformly in $K(z_0; r)$. It follows by Hurwitz's theorem (§1.4) that $f'(z)f(z)^n \equiv 1$ in $\mathbb{C}$. Therefore,

$$f(z)^{n+1} = (n+1)z + c \qquad (c \text{ constant}),$$

which contradicts the fact that f is meromorphic in $\mathbb{C}$. We conclude that $\langle f, \mathbb{C} \rangle \in \mathcal{P}$, establishing (iii). Condition (iv) follows from the results of Hayman/Mues cited above, and so $\mathcal{F}$ is normal by the Heuristic Principle.

Note. Interestingly enough, if Hayman's conjecture: $f'f \neq 1$ *for* f *meromorphic in* $\mathbb{C}$ *implies* $f \equiv$ *constant*, is true, Theorem 4.5.2 holds for $n = 1$ by setting $k = \frac{1}{2}$ in the preceding proof. A further consequence of the truth of the conjecture would be a proof of the meromorphic version of Theorem 4.4.19 for $\nu = 3$ (cf. the proof of Theorem 3 in Pang [1990]).

Another generalization of this type comes from Schwick [1989].

Theorem 4.5.3 *Let* $\mathcal{F}$ *be a family of analytic (meromorphic) functions on a domain* Ω *satisfying the condition*

$$(f^n)^{(k)}(z) \neq 1, \qquad z \in \Omega,$$

where $n \geq k + 1$ $(n \geq k + 3)$. *Then* $\mathcal{F}$ *is normal in* Ω.

The meromorphic case has been extended to $n \geq k + 2$ Hua [preprint]. It is worth noting that an analytic (meromorphic) function satisfying the conditions of the theorem in $\mathbb{C}$ is necessarily constant (Clunie [1967]) for the case $k = 1$; Hennekemper [1979] for the general case).

We may further deduce from Theorem 4.5.1, for $q = 1$, $s > 1$,

Corollary 4.5.4 *If $\mathcal{F}$ is a family of meromorphic functions in Ω such that $f(z) \neq 0$, $f^{(k)}(z) \neq 1$, and no poles are simple, for each $f \in \mathcal{F}$, then $\mathcal{F}$ is normal.*

The restriction in this corollary on the poles not being simple is actually unnecessary, and the result was proved without this restriction by Ku [1979]. We present here the proof given by Yang [1982]. What is proved is the following, from which the latter Ku result is a direct consequence.

Theorem 4.5.5 *Given $k \in \mathbb{N}$, and $f(z)$ meromorphic in U such that $f(z) \neq 0$, $f^{(k)}(z) \neq 1$, then either $|f(z)| < 1$ or $|f(z)| > C$, for all $|z| < \frac{1}{32}$, where C depends only on k.*

As usual, a number of preliminary lemmas must first be established for the proof; k will always be a fixed positive integer. The first result is essentially due to Milloux [1940], and is analogous to the basic inequality (4.8) of Drasin.

Lemma 4.5.6 *Let $f(z)$ be meromorphic in $|z| < \rho \leq \infty$, $f(0) \neq 0, \infty$, $f^{(k)}(0) \neq 1$ and $f^{(k+1)}(0) \neq 0$. Then*

$$T(r, f)$$
$$< \bar{N}(r, f) + N\left(r, \frac{1}{f}\right) + N\left(r, \frac{1}{f^{(k)} - 1}\right) - N\left(r, \frac{1}{f^{(k+1)}}\right) + S(r, f),$$
$$0 < r < \rho,$$

with

$$S(r, f) = m\left(r, \frac{f^{(k)}}{f}\right) + m\left(r, \frac{f^{(k+1)}}{f}\right) + m\left(r, \frac{f^{(k+1)}}{f^{(k)} - 1}\right)$$
$$+ \log \left| \frac{f(0)\left\{f^{(k)}(0) - 1\right\}}{f^{(k+1)}(0)} \right| + \log 2. \qquad (4.66)$$

Proof. Starting with the identity

$$\frac{1}{f} = \frac{f^{(k)}}{f} - \frac{f^{(k)} - 1}{f^{(k+1)}} \cdot \frac{f^{(k+1)}}{f},$$

we obtain

$$m\left(r, \frac{1}{f}\right) \leq m\left(r, \frac{f^{(k)}}{f}\right) + m\left(r, \frac{f^{(k)} - 1}{f^{(k+1)}}\right) + m\left(r, \frac{f^{(k+1)}}{f}\right) + \log 2.$$

From the Jensen formula (1.6) applied to $m\left(r,\frac{1}{f}\right)$, and then in the form

$$m\left(r,\frac{g}{g'}\right) = m\left(r,\frac{g'}{g}\right) + N\left(r,\frac{g'}{g}\right) - N\left(r,\frac{g}{g'}\right) + \log\left|\frac{g(0)}{g'(0)}\right|,$$

applied to $m\left(r,\frac{f^{(k)}-1}{f^{(k+1)}}\right)$ with $g = f^{(k)} - 1$, we arrive at

$$T(r,f) \le N\left(r,\frac{1}{f}\right) + N\left(r,\frac{f^{(k+1)}}{f^{(k)}-1}\right) - N\left(r,\frac{f^{(k)}-1}{f^{(k+1)}}\right) + S(r,f),$$

for $S(r,f)$ given by (4.66). Now

$$N\left(r,\frac{f^{(k+1)}}{f^{(k)}-1}\right) - N\left(r,\frac{f^{(k)}-1}{f^{(k+1)}}\right)$$
$$= \bar{N}(r,f) + N\left(\frac{1}{f^{(k)}-1}\right) - N\left(r,\frac{1}{f^{(k+1)}}\right),$$

which proves the lemma.

The next requirement is essentially Theorem 1 of Hayman [1959] (cf. also Theorem 3.5 of Hayman [1964]), except that the error term $S_1(r,f)$ of that theorem, according to Lemma 4.5.6 can be expressed by (4.66).

Lemma 4.5.7 *Let $f(z)$ be meromorphic in U such that $f(0) \ne 0,\infty$, $f^{(k)}(0) \ne 1$, $f^{(k+1)}(0) \ne 0$, and*

$$(k+1)f^{(k+2)}(0)\{f^{(k)}(0) - 1\} - (k+2)\{f^{(k+1)}(0)\}^2 \ne 0.$$

Then

$$T(r,f) < \left(2+\frac{1}{k}\right)N\left(r,\frac{1}{f}\right) + \left(2+\frac{2}{k}\right)\bar{N}\left(\frac{1}{f^{(k)}-1}\right) + S(r,f),$$
$$0 < r < 1, \quad (4.67)$$

where

$$S(r,f)$$
$$= \left(2+\frac{2}{k}\right)m\left(r,\frac{f^{(k+1)}}{f^{(k)}-1}\right) + \left(2+\frac{1}{k}\right)\left\{m\left(r,\frac{f^{(k+1)}}{f}\right) + m\left(r,\frac{f^{(k)}}{f}\right)\right\}$$
$$+ \frac{1}{k}m\left(r,\frac{f^{(k+2)}}{f^{(k+1)}}\right) + \left(2+\frac{1}{k}\right)\log\left|\frac{f(0)\left\{f^{(k)}(0)-1\right\}}{f^{(k+1)}(0)}\right| + 4$$
$$+ \frac{1}{k}\log\left|\frac{f^{(k+1)}(0)\left\{f^{(k)}(0)-1\right\}}{(k+1)f^{(k+2)}(0)\{f^{(k)}(0)-1\}-(k+2)\{f^{(k+1)}(0)\}^2}\right|. \quad (4.68)$$

Our last lemma is

Lemma 4.5.8 *Suppose that $f(z)$ satisfies the assumptions of Lemma 4.5.6, as well as $f(z) \neq 0$, $f^{(k)}(z) \neq 1$ in U. Then given $\rho < 1$ and $0 < r < \rho$,*

$$\log M\left(r, \frac{1}{f}\right) < \frac{C}{\rho - r}\left(1 + B + \log\frac{1}{\rho - r}\right),$$

where

$$B = \log^+ |f(0)| + \log^+ |f^{(k)}(0)| + \log^+ \frac{1}{|f^{(k+1)}(0)|}$$

$$+ \log^+ \frac{1}{|(k+1)f^{(k+2)}(0)\{f^{(k)}(0) - 1\} - (k+2)\{f^{(k+1)}(0)\}^2|}. \quad (4.69)$$

Proof. From the additional assumptions we get $T(r, f) < S(r, f)$ in (4.67) and (4.68). We must first estimate the terms of (4.68).

According to the Nevanlinna estimate (Lemma 4.4.3) for $0 < \delta < r < R = \frac{r+\rho}{2} < \rho < 1$,

$$m\left(r, \frac{f^{(k+1)}}{f^{(k)} - 1}\right)$$

$$< C\left\{\log^+ T(R, f^{(k)}) + \log\frac{1}{R - r} + \log^+\log^+ \frac{1}{|f^{(k)}(0) - 1|} + 1\right\}, \quad (4.70)$$

and

$$m\left(r, \frac{f^{(k+2)}}{f^{(k+1)}}\right)$$

$$< C\left\{\log^+ T(R, f^{(k+1)}) + \log\frac{1}{R - r} + \log^+\log^+ \frac{1}{|f^{(k+1)}(0)|} + 1\right\}. \quad (4.71)$$

To treat the terms $\log^+ T(R, f^{(j)})$, $j = k, k+1$, in (4.70) and (4.71), we note that

$$\log^+ T(R, f^{(j)}) \leq \log^+\left\{(j+1)T(R, f) + m\left(R, \frac{f^{(j)}}{f}\right)\right\}$$

$$< \log^+ T(R, f) + m\left(R, \frac{f^{(j)}}{f}\right) + C.$$

We conclude from (4.67), (4.68), (4.70), and (4.71) that

$$T(r, f) < C\left\{1 + \log^+ T(R, f) + \log\frac{1}{R - r}\right.$$

$$+ \log^+\log^+ \frac{1}{|f^{(k)}(0) - 1|} + \log^+\log^+ \frac{1}{|f^{(k+1)}(0)|}\right\}$$

$$+ \left(2 + \frac{1}{k}\right)\log|f(0)| + \left(2 + \frac{2}{k}\right)\log|f^{(k)}(0) - 1| + 2\log\frac{1}{|f^{(k+1)}(0)|}$$

$$+ \frac{1}{k} \log \frac{1}{|(k+1)f^{(k+2)}(0)\left\{f^{(k)}(0)-1\right\} - (k+2)\left\{f^{(k+1)}(0)\right\}^2|}$$

$$+ C_1 \left\{ m\left(R, \frac{f^{(k)}}{f}\right) + m\left(R, \frac{f^{(k+1)}}{f}\right) \right\}. \tag{4.72}$$

We next apply the Hiong estimate (Lemma 4.4.3) to the last two terms of (4.72), with r, R replaced by R, ρ, respectively. As a consequence, in view of the relationship between r, R, and ρ,

$$T\left(r, \frac{1}{f}\right) < C\left\{ 1 + \log \frac{1}{R-r} + \log^+ T(R,f) \right.$$

$$+ \log^+ \log^+ \frac{1}{|f(0)|} + \log^+ \log^+ \frac{1}{|f^{(k)}(0)-1|} + \log^+ \log^+ \frac{1}{|f^{(k+1)}(0)|} \Big\}$$

$$+ \left(1 + \frac{1}{k}\right) \log|f(0)| + \left(2 + \frac{2}{k}\right) \log|f^{(k)}(0) - 1| + 2\log \frac{1}{|f^{(k+1)}(0)|}$$

$$+ \frac{1}{k} \log \frac{1}{|(k+1)f^{(k+2)}(0)\left\{f^{(k)}(0)-1\right\} - (k+2)\left\{f^{(k+1)}(0)\right\}^2|}. \tag{4.73}$$

Letting $x = |f(0)|$ and subsequently $x = |f^{(k)}(0) - 1|$ in (4.10) and applying the inequality

$$\log^+ T(R,f)$$

$$= \log^+ \left\{ T\left(R, \frac{1}{f}\right) + \log|f(0)| \right\} \le \log^+ T\left(R, \frac{1}{f}\right) + \log^+ |f(0)| + 1,$$

(4.73) becomes

$$T\left(r, \frac{1}{f}\right) < C_1[1+B] + C_2 \left\{ \log \frac{1}{R-r} + \log^+ T\left(R, \frac{1}{f}\right) \right\} \tag{4.74}$$

where B is given by (4.69). Taking $C_2 > 1$ in Drasin's Lemma 4.4.4, then (4.74) yields

$$T\left(r, \frac{1}{f}\right) < C\left(1 + B + \log \frac{1}{\rho - r}\right), \qquad \frac{\rho}{2} = \delta < r < \rho < 1. \tag{4.75}$$

Since $\frac{1}{f}$ is analytic in U,

$$\log M\left(r, \frac{1}{f}\right) \le \frac{\rho + 3r}{\rho - r} T\left(\frac{r+\rho}{2}, \frac{1}{f}\right), \qquad 0 < r < \rho, \tag{4.76}$$

so that the result follows by an application of (4.75).

We are now ready to prove Theorem 4.5.5. The theorem obtains with $C = 1$ unless there is some point z_1 with

$$|f(z_1)| = 1, \qquad |z_1| < \tfrac{1}{32}.$$

Two mutually exclusive cases are considered under the foregoing assumption and we demonstrate that $|f(z)| > C$ for all $|z| < \frac{1}{32}$, in each case.

Case (i). Suppose that

$$\sum_{j=0}^{k+1} |f^{(j)}(z)| \geq \frac{1}{4}, \qquad |z| < \frac{1}{8}.$$

Then

$$\frac{1}{|f(z)|} \leq 4 \sum_{j=0}^{k+1} \left| \frac{f^{(j)}(z)}{f(z)} \right|, \qquad |z| < \frac{1}{8}.$$

Denote by $m(r, z_1, f)$, $M(r, z_1, f)$ and $T(r, z_1, f)$, the quantities $m(r, f(z_1 + z))$, $M(r, f(z_1 + z))$, $T(r, f(z_1 + z))$, respectively, with $|z| = r$, so that

$$m\left(r, z_1, \frac{1}{f}\right) \leq \sum_{j=0}^{k+1} m\left(r, z_1, \frac{f^{(j)}}{f}\right) + \log 4(k+2), \qquad 0 < r < \tfrac{3}{32}. \quad (4.77)$$

The Hiong estimate is applicable to $f(z_1 + z)$, and since $N\left(r, z_1, \frac{1}{f}\right) = 0$, we obtain from (4.77)

$$T\left(r, z_1, \frac{1}{f}\right) = m\left(r, z_1, \frac{1}{f}\right) \leq C\left(1 + \log \frac{1}{R-r} + \log^+ T(R, z_1, f)\right),$$
$$\tfrac{1}{32} \leq r < R \leq \tfrac{3}{32}. \quad (4.78)$$

By Jensen's formula (1.6), since $|f(z_1)| = 1$, the term $\log^+ T(R, z_1, f)$ may be replaced by $\log^+ T\left(R, z_1, \frac{1}{f}\right)$ in (4.78). In view of Lemma 4.4.4,

$$T\left(r, z_1, \frac{1}{f}\right) < C\left(1 + \log \frac{1}{R-r}\right), \qquad \tfrac{1}{32} \leq r < R \leq \tfrac{3}{32},$$

implying $T\left(\frac{5}{64}, z_1, \frac{1}{f}\right) < C$. We conclude, invoking (4.76), that

$$\log M\left(\frac{1}{32}, \frac{1}{f}\right) \leq \log M\left(\frac{1}{16}, z_1, \frac{1}{f}\right) \leq 9T\left(\frac{5}{64}, z_1, \frac{1}{f}\right) < C,$$

and the assertion follows in this case.

Case (ii). There is a point z_2 such that

$$\sum_{j=0}^{k+1} |f^{(j)}(z_2)| < \frac{1}{4}, \qquad |z_2| < \tfrac{1}{8}. \quad (4.79)$$

We claim there is some point z_0 on the line segment $\overline{z_1 z_2}$ with

$$|f^{(k+2)}(z_0)| \geq 1,$$

$$\frac{1}{12} < |f^{(k+1)}(z_0)| < \frac{1}{2}, \quad |f^{(k)}(z_0)| < \frac{1}{2}, \quad |f(z_0)| < \frac{1}{2}. \quad (4.80)$$

To see this, first note that if $|f^{(k+1)}(z)| < \frac{1}{4}$ for z on $\overline{z_1 z_2}$, we deduce from (4.79) that

$$|f^{(k)}(z)| \leq |f^{(k)}(z_2)| + \left| \int_{\overline{z_2 z}} f^{(k+1)}(\zeta)\, d\zeta \right| < \frac{1}{4} + \frac{1}{4}|z_2 - z| < \frac{1}{3},$$

and by successive applications of this argument,

$$|f^{(j)}(z)| < \frac{1}{3}, \qquad j = k-1, \ldots, 1, 0.$$

However, for $j = 0$, $|f(z_1)| < \frac{1}{3}$, contradicting our assumption $|f(z_1)| = 1$. Hence there is a point z_3 on $\overline{z_1 z_2}$ with $|f^{(k+1)}(z_3)| = \frac{1}{4}$ and $|f^{(k+1)}(z)| < \frac{1}{4}$ on $\overline{z_2 z_3}$, which leads to

$$|f^{(k)}(z_3)| \leq |f^{(k)}(z_2)| + \left| \int_{\overline{z_2 z_3}} f^{(k+1)}(\zeta)\, d\zeta \right| < \frac{1}{3}.$$

As before

$$|f^{(j)}(z_3)| < \frac{1}{3}, \qquad j = k-1, \ldots, 1, 0,$$

obtains. Thus we may take $z_3 = z_0$ in (4.80) if $|f^{(k+2)}(z_3)| \geq 1$.

Suppose then $|f^{(k+2)}(z_3)| < 1$. If $|f^{(k+2)}(z)| < 1$ on $\overline{z_1 z_3}$, then for z on $\overline{z_1 z_2}$,

$$|f^{(k+1)}(z)| < \frac{1}{4} + \frac{1}{8} + \frac{1}{32} < \frac{1}{2},$$

implying

$$|f^{(k)}(z)| \leq |f^{(k)}(z_2)| + |z_2 - z_1| \cdot \max |f^{(k+1)}(z)| < \frac{1}{3}.$$

Therefore, as above, $|f^{(j)}(z)| < \frac{1}{3}$ for z on $\overline{z_1 z_2}$, $j = 0, 1, \ldots, k$, again contradicting $|f(z_1)| = 1$. Thus there must be a point z_4 on $\overline{z_1 z_3}$ with $|f^{(k+2)}(z_4)| = 1$ and $|f^{(k+2)}(z)| < 1$ on $\overline{z_3 z_4}$. Note that $|z_3 - z_4| < |z_1 - z_2| < \frac{1}{32} + \frac{1}{8} = \frac{5}{32}$ so that for z on $\overline{z_3 z_4}$,

$$|f^{(k+1)}(z)| \geq |f^{(k+1)}(z_3)| - |z_3 - z_4| \cdot \max |f^{(k+2)}(z)| > \frac{1}{12},$$

$$|f^{(k+1)}(z)| \leq |f^{(k+1)}(z_3)| + |z_3 - z_4| \cdot \max |f^{(k+2)}(z)| < \frac{1}{2}.$$

It follows that

$$|f^{(k)}(z_4)| \leq |f^{(k)}(z_3)| + |z_3 - z_4| \cdot \max |f^{(k+1)}(z)| < \frac{1}{2},$$

and likewise

$$|f^{(j)}(z_4)| < \frac{1}{2}, \qquad j = k-1, \ldots, 1, 0,$$

so that in this instance we may take $z_0 = z_4$ in (4.80), and (4.80) holds in all cases.

We now wish to apply Lemma 4.5.8 to $f(z)$ in $|z - z_0| < \frac{7}{8}$. From (4.80)

$$|(k+1)f^{(k+2)}(z_0)\{f^{(k)}(z_0) - 1\} - (k+2)\{f^{(k+1)}(z_0)\}^2|$$
$$> \frac{k+1}{2} - \frac{k+2}{4} \geq \frac{1}{4},$$

and consequently by Lemma 4.5.8

$$\log M\left(\frac{1}{2}, z_0, \frac{1}{f}\right) < C.$$

Finally

$$\log M\left(\frac{1}{32}, \frac{1}{f}\right) < \log M\left(\frac{1}{2}, z_0, \frac{1}{f}\right) < C,$$

completing the proof of the theorem.

In view of the local nature of the foregoing analysis, we may conclude

Corollary 4.5.9 *If $\mathcal{F}$ is a family of meromorphic functions in Ω such that for some fixed $k \in \mathbb{N}$, $f(z) \neq 0$, $f^{(k)}(z) \neq 1$, for each $f \in \mathcal{F}$, then $\mathcal{F}$ is normal in Ω.*

See Hiong and Ho [1961] for a related analytic result.

Remark. The technique employed in Case (ii) was also utilized in Yang and Chang [1966]. The proof of the theorem captures the spirit of the "Chinese School" approach, and it has been very successful in dealing with the normality of families of meromorphic functions. This approach was also adopted by Langley [1984] in his proof of the meromorphic version of Theorem 4.4.19. Furthermore, using techniques similar to those in the proof of the preceding theorem, Yang [1986a] has given a criterion for normality in terms of fixed points. Recall, a point z_0 is a *fixed point* of a function $f(z)$ if $f(z_0) = z_0$. Yang demonstrated

Theorem 4.5.10 *Let $\mathcal{F}$ be a family of meromorphic functions on a domain Ω, $k \in \mathbb{N}$. If for each $f \in \mathcal{F}$, both $f(z)$ and $f^{(k)}(z)$ have no fixed points in Ω, then $\mathcal{F}$ is normal.*

In a similar vein, Yang [1986b] has given the following generalization of Corollary 4.5.9.

Theorem 4.5.11 *Let $\mathcal{F}$ be a family of meromorphic functions in Ω, $k \in$ $\mathbb{N}$. Suppose that $\phi(z)$ and $\psi(z)$ are analytic in Ω and $\phi^{(k)}(z) \not\equiv \psi(z)$. If for every $f \in \mathcal{F}$, $f(z) \neq \phi(z)$ and $f^{(k)}(z) \neq \psi(z)$, $z \in \Omega$, then $\mathcal{F}$ is normal in Ω.*

The final result in this cycle is Yang's [1985] extension of Theorem 4.4.14.

Theorem 4.5.12 *Let Ω be a domain, $k \geq 2$ a positive integer, and $a_j(z)$, $j = 0, 1, \ldots, k-2$, analytic functions in Ω. If $\mathcal{F}$ is a family of meromorphic functions such that for $f \in \mathcal{F}$, $f(z) \neq 0$ and*

$$f^{(k)}(z) + \sum_{j=0}^{k-2} a_j(z) f^{(j)}(z) \neq 1, \qquad z \in \Omega,$$

then $\mathcal{F}$ is normal.

A Modified Drasin/Zalcman Approach. Combining the Zalcman Lemma (§4.1) with the Nevanlinna theory creates another powerful approach with which to investigate the nature of normal families. This technique first appeared in Oshkin [1982], coupled with some parts of the Drasin theory, and subsequently in S. Li [1984]. Moreover, Schwick [1989] has also grafted on some of the Drasin theory to produce several new results, one of which we examine in detail below.

Consider a family $\mathcal{F}$ of meromorphic functions on $\Omega \subseteq \mathbb{C}$ satisfying the condition

$$f^{(\ell)}(z) \neq 0, \qquad \ell = 0, 1, 2, \ldots, \ z \in \Omega. \tag{4.81}$$

When $\Omega = \mathbb{C}$, then any $f \in \mathcal{F}$ must be of the form (Hayman [1959]),

$$f_1(z) = e^{az+b} \quad \text{or} \quad f_2(z) = \frac{1}{(Az + B)^k},$$

$a \neq 0$, $A \neq 0$, $k \in \mathbb{N}$. In view of the examples $\{e^{nz}\}$ and $\{\frac{1}{nz}\}$, families satisfying (4.81) would not, in general, be normal. On the other hand, Schwick proves

Theorem 4.5.13 *For $\mathcal{F}$ given as in (4.81), the associated family*

$$\tilde{\mathcal{F}} = \left\{ \frac{f'}{f} : f \in \mathcal{F} \right\}$$

is normal in Ω.

Before proving the theorem we require a few preliminary results, one of which is a local adaptation of the

Zalcman Lemma *A family $\mathcal{F}$ of meromorphic (analytic) functions in U is not normal at the origin if and only if there exists a sequence $\{f_n\} \subseteq \mathcal{F}$, a sequence $z_n \to 0$, a positive sequence $\rho_n \to 0$, and a nonconstant meromorphic (entire) function $g(z)$ on $\mathbb{C}$ such that*

$$\lim_{n \to \infty} f_n(z_n + \rho_n z) = g(z)$$

normally on $\mathbb{C}$.

The proof is an obvious modification of the original (§4.1).

A further elementary estimate is required (Schwick [1989]).

Lemma 4.5.14 *Let $f(z)$ be meromorphic in $|z| < \rho \leq \infty$, $k \in \mathbb{N}$, with $f(0) \neq \infty$. Then*

$$m(r, f^{(k)})$$
$$< m(r, f) + C \left\{ \log^+ T(R, f) + \log^+ \frac{1}{R - r} + \log^+ \frac{1}{r} + \log^+ R + 1 \right\},$$
$$0 < r < R < \rho,$$

and C depends only on k.

Proof. Assuming that $|f(0)| \leq 1$, then from

$$m(r, f^{(k)}) \leq m(r, f + 2) + m\left(r, \frac{f^{(k)}}{f + 2}\right)$$

and the Hiong estimate (Lemma 4.4.3), the conclusion follows. For $|f(0)| > 1$ and

$$m(r, f^{(k)}) \leq m(r, f) + m\left(r, \frac{f^{(k)}}{f}\right)$$

together with the Hiong estimate, the desired inequality again holds.

The interesting feature of this lemma is that the estimate does not depend on the initial-value term containing $f(0)$.

Our next result is due to Pólya [1922] (cf. Hayman [1964], Theorem 3.6).

Lemma 4.5.15 *If $f(z)$ is meromorphic in $|z - z_0| < R \leq \infty$, analytic in $|z - z_0| < r < R$, and has at least two poles on $|z - z_0| = r$, then there exists $\delta > 0$ and $\ell_0 = \ell_0(\delta)$ such that $f^{(\ell)}$ has a zero in $|z - z_0| < \delta$ for $\ell \geq \ell_0$.*

Consequently we have

Lemma 4.5.16 *Let $f(z)$ be meromorphic in U and satisfy $f^{(\ell)}(z) \neq 0$ for all $\ell \in \mathbb{N}$, $z \in U$. Then f has at most one pole in $|z| < \frac{1}{4}$.*

Proof. Suppose that f has at least two poles in $|z| < \frac{1}{4}$, and denote one of them by z_0. Define

$$(0 <)m = \min\{|z_p - z_0| : z_p \text{ is a pole of } f, z_p \neq z_0\} < \frac{1}{2}.$$

For z_m a pole of f satisfying $|z_m - z_0| = m$, the disk

$$K = \left\{ z : \left| z - \frac{z_0 + z_m}{2} \right| \leq \frac{m}{2} \right\}$$

is contained in U. Moreover, f is analytic in K^0 and has exactly two poles on ∂K, z_0 and z_m. By Lemma 4.5.15 we obtain a contradiction with the hypothesis.

Lemma 4.5.17 *A family $\mathcal{F}$ of linear functions is not normal at the origin if and only if there is a sequence $\{f_n = a_n z + b_n\} \subseteq \mathcal{F}$ satisfying* (i) $a_n \to \infty$ *and* (ii) $b_n/a_n \to 0$.

Proof. If $\mathcal{F}$ is not normal at the origin, then by the Zalcman Lemma above, there is a sequence $\{f_n\} \subseteq \mathcal{F}$, $z_n \to 0$, $\rho_n \to 0^+$ and a nonconstant entire function g with

$$f_n(z_n + \rho_n z) = a_n \rho_n z + a_n z_n + b_n \to g(z)$$

normally in $\mathbb{C}$. But then $g(z) = az + b$ $(a \neq 0)$ by the Weierstrass Theorem, implying $a_n \to \infty$. Furthermore, the zeros of $f_n(z_n + \rho_n z)$ converge to the sole zero of $g(z)$, namely, $-b/a$, that is to say

$$-\frac{1}{\rho_n}\left(z_n + \frac{b_n}{a_n} \right) \to -\frac{b}{a},$$

proving (ii).

Conversely, suppose $\{f_n\} \subseteq \mathcal{F}$ satisfies (i) and (ii), and set $\rho_n = 1/|a_n|$, $z_n = -b_n/a_n$. Then

$$f_n(z_n + \rho_n z) = a_n(z_n + \rho_n z) + b_n = \frac{a_n}{|a_n|} z = c_n z, \qquad |c_n| = 1.$$

Then a subsequence of $f_n(z_n + \rho_n z)$ tends normally to an entire function of the form $g(z) = cz$, $|c| = 1$, and so the converse follows from the Zalcman Lemma.

In the sequel denote

$$g(z) = g(f)(z) = \frac{f(z)}{f'(z)}.$$

Note that under the assumption (4.81), g is analytic, having simple zeros at the poles of f. Then we can show

Theorem 4.5.18 *Let $\{f_n\}$ be a sequence of meromorphic functions on U such that each f_n satisfies (4.81) and $g_n'' \equiv 0$ $(g_n = g(f_n))$. Then $\{g_n\}$ is normal in a neighbourhood of the origin.*

Proof. Suppose that $\{g_n\}$ is not normal at the origin. By the assumption on g_n, we have $g_n = a_n z + b_n$; whence we may suppose that (i) $a_n \to \infty$ and (ii) $b_n/a_n \to 0$ by Lemma 4.5.17 . Note

$$\frac{f_n'(z)}{f_n(z)} = \frac{1}{a_n z + b_n} = \frac{1}{a_n \left(z + \frac{b_n}{a_n} \right)},$$

and so f_n has a pole at $-b_n/a_n$. Then for each fixed n, computing $\frac{1}{2\pi i} \int_{C_n} \frac{f_n'}{f_n}$ over a sufficiently small circle C_n gives $-\frac{1}{a_n} \in \mathbb{N}$, by virtue of the Argument Principle (cf. Palka [1991], p. 340). This contradicts (i).

We, therefore, make the assumption

$$g'' = \left(\frac{f}{f'} \right)'' \not\equiv 0,$$

throughout the sequel.

Theorem 4.5.19 *Let f be meromorphic in U satisfying (4.81) and suppose that $g(0) \neq 0, \infty$, $g'(0) \neq 1$, $g''(0) \neq 0$, $(g = g(f))$. Then*

$$T(r, g) \;<\; \bar{N}(r, f) + \log^+ |g(0)| + \log^+ |g'(0) - 1| + \log \left| \frac{1}{g''(0)} \right|$$

$$+ C \left\{ \log^+ T(R, g) + \log \frac{1}{R - r} + \log^+ \frac{1}{r} + 1 \right\},$$

$0 < r < R < 1$, where C is independent of f.

Proof. Since $g' = 1 - f f''/f'^2$, we see that $N\left(r, \frac{1}{g'-1} \right) = 0$. From the Milloux estimate (Lemma 4.5.6) for g, with $k = 1$,

$$T(r, g) \;<\; \bar{N}(r, g) + N\left(r, \frac{1}{g} \right) + N\left(r, \frac{1}{g'-1} \right)$$

$$- N\left(r, \frac{1}{g''} \right) + \log \left| \frac{g(0)[g'(0) - 1]}{g''(0)} \right| + m\left(r, \frac{g'}{g} \right) + m\left(r, \frac{g''}{g} \right)$$

$$+ m\left(r, \frac{g''}{g'-1} \right) + A.$$

Noting that $\bar{N}(r,g) = 0$ and $N\left(r, \frac{1}{g}\right) = \bar{N}(r,f)$, an application of the Hiong estimate (Lemma 4.4.3) to the preceding inequality yields for $0 < r < r_1 = \frac{r+R}{2} < R < 1$

$$T(r,g) < \bar{N}(r,f) + \log\left|\frac{g(0)[g'(0) - 1]}{g''(0)}\right|$$
$$+C\left\{\log^+ T(R,g) + \log^+\log^+\frac{1}{|g(0)|} + \log^+ T(r_1, g' - 1)\right.$$
$$\left. + \log^+\log^+\frac{1}{|g'(0) - 1|} + \log\frac{1}{r_1 - r} + \log\frac{1}{R - r} + \log^+\frac{1}{r} + 1\right\}. \quad (4.82)$$

To treat the term $\log^+ T(r_1, g' - 1)$ in (4.82) we can apply Lemma 4.5.14 which gives, upon simplification,

$$T(r,g) \quad < \quad \bar{N}(r,f) + \log\left|\frac{g(0)[g'(0) - 1]}{g''(0)}\right|$$
$$+C\left\{\log^+ T(R,g) + \log^+\log^+\frac{1}{|g(0)|} + \log^+\log^+\frac{1}{|g'(0) - 1|}\right.$$
$$\left. + \log\frac{1}{R - r} + \log^+\frac{1}{r} + 1\right\}. \quad (4.83)$$

Then the theorem follows by an application of (4.10).

We are now ready to prove Theorem 4.5.13. In fact, let

$$\widetilde{\mathcal{G}} = \left\{g = \frac{f}{f'} : f \in \mathcal{F}\right\}, \qquad \Omega = U,$$

and note that $\widetilde{\mathcal{F}}$ is normal if and only if $\widetilde{\mathcal{G}}$ is normal. We prove that $\widetilde{\mathcal{G}}$ is normal at the origin by assuming the contrary. Then by the Zalcman Lemma, there exists a sequence $\{g_n\} \subseteq \widetilde{\mathcal{G}}$, $z_n \to 0$, $\rho_n \to 0$, and a nonconstant entire function $g(z)$ with

$$\lim_{n\to\infty} g_n(z_n + \rho_n z) = g(z)$$

normally in $\mathbb{C}$. Let $\{f_n\}$ denote the corresponding sequence in $\mathcal{F}$. The argument splits into two cases, in each of which our initial aim is to demonstrate the boundedness of the terms

$$\bar{N}(r, f_n) + \log^+|g_n(0)| + \log^+|g_n'(0) - 1| + \log\left|\frac{1}{g_n''(0)}\right|,$$

when applying the previous theorem to $\{g_n\}$.

Case 1. Suppose that every f_n is analytic in $|z| < r_1$, for $0 < r_1 < 1$. Then each g_n is zero-free analytic in $|z| < r_1$, and thus g is zero-free, so that

g cannot be linear. Choose $z_0 \in \mathbb{C}$ such that $(g(z_0) \neq 0, \infty)$, $g'(z_0) \neq 0$, and $g''(z_0) \neq 0$. Then

$$\log^+ |g_n(z_n + \rho_n z_0)| + \log^+ |g'_n(z_n + \rho_n z_0) - 1| + \log \left| \frac{1}{g''_n(z_n + \rho_n z_0)} \right| + \log \frac{1}{\rho_n}$$

$$= \log^+ |g_n(z_n + \rho_n z_0)| + \log^+ \left| \frac{1}{\rho_n} \{ g_n(z_n + \rho_n z)'(z_0) - \rho_n \} \right|$$

$$+ \log \left| \frac{1}{g_n(z_n + \rho_n z)''(z_0)} \right| + \log \rho_n^2 + \log \frac{1}{\rho_n}$$

$$\rightarrow \log^+ |g(z_0)| + \log |g'(z_0)| + \log \left| \frac{1}{g''(z_0)} \right|, \qquad \text{as } n \to \infty.$$

This shows that

$$\log^+ |g_n(z_n + \rho_n z_0)| + \log^+ |g'_n(z_n + \rho_n z_0) - 1| + \log \left| \frac{1}{g''(z_n + \rho_n z_0)} \right|$$

$\rightarrow -\infty$ as $n \to \infty$. Without loss of generality let $z_n + \rho_n z_0 = 0$. Then for all n sufficiently large, the hypotheses of Theorem 4.5.19 are satisfied,

$$\bar{N}(r, f_n) = 0, \qquad 0 < r < r_1,$$

and therefore

$$T(r, g_n) < C \left\{ \log^+ T(R, g_n) + \log \frac{1}{R - r} + 1 \right\}, \qquad \tfrac{1}{2} r_1 < r < R < r_1.$$

The $\log^+$ term is dispensed with as usual by Lemma 4.4.4, and the result follows by virtue of Theorem 3.6.9.

Case 2. Suppose that f_n has a pole at a_n with $a_n \to 0$, passing to a subsequence if necessary. We again claim that g is not a linear function. For, assume that

$$g(z) = az + b, \qquad a \neq 0,$$

so that g has a zero at $-b/a$. Then by the Hurwitz Theorem there exists $\tilde{z}_n \to -b/a$ with

$$g_n(z_n + \rho_n \tilde{z}_n) = 0$$

for all n sufficiently large. As the (simple) zeros of g_n are the poles of f_n, and f_n has but a single pole in $|z| < \frac{1}{4}$ (Lemma 4.5.16), namely a_n, it follows that $z_n + \rho_n \tilde{z}_n = a_n$. Letting m_n denote the order of this pole,

$$g'_n(z_n + \rho_n \tilde{z}_n) = g'_n(a_n) = -\frac{1}{m_n}$$

and hence

$$g_n(z_n + \rho_n z)'(\tilde{z}_n) = \rho_n g'_n(z_n + \rho_n \tilde{z}_n) = -\rho_n/m_n \to 0.$$

On the other hand,

$$g(z_n + \rho_n z)'(\widetilde{z}_n) \to g'(-b/a) = a \neq 0,$$

from which we conclude that $g(z)$ is not linear.

As above, choose $z_0 \in \mathbb{C}$ with $g(z_0) \neq 0, \infty$, $g'(z_0) \neq 0, \infty$, and $g''(z_0) \neq 0, \infty$, which again allows us to deduce the boundedness of

$$\log^+ |g_n(z_n + \rho_n z_0)| + \log^+ |g'_n(z_n + \rho_n z_0) - 1| + \log \left| \frac{1}{g''_n(z_n + \rho_n z_0)} \right| + \log \frac{1}{\rho_n}.$$

We now require a bound for $\bar{N}(r, f_n)$.

(a) If g is zero-free, then for $|z - z_0| < 1$ and large n

$$g_n\big(z_n + \rho_n z_0 + \rho_n(z - z_0)\big) = g_n(z_n + \rho_n z) \neq 0,$$

and therefore

$$|a_n - (z_n + \rho_n z_0)| \geq \rho_n.$$

(b) If $g(w) = 0$, then there exists $\widetilde{z}_n$ satisfying

$$\widetilde{z}_n \to w, \quad g_n(z_n + \rho_n \widetilde{z}_n) = 0.$$

As above, $z_n + \rho_n \widetilde{z}_n = a_n$ and

$$\begin{aligned} |a_n - (z_n + \rho_n z_0)| &= |z_n + \rho_n \widetilde{z}_n - z_n - \rho_n z_0| \\ &= \rho_n |\widetilde{z}_n - z_0| \geq \rho_n \frac{|w - z_0|}{2}. \end{aligned}$$

The estimates of both (a) and (b) can be reformulated as

$$|a_n - (z_n + \rho_n z_0)| > C\rho_n, \qquad C \neq 0,$$

for all large n. Again, let us assume that $z_n + \rho_n z_0 = 0$. For $0 < r < \frac{1}{4}$,

$$\bar{N}(r, f_n) = \log \frac{r}{|a_n|} \leq \log \frac{r}{C\rho_n},$$

implying the boundedness of

$$\bar{N}(r, f_n) + \log^+ |g_n(0)| + \log^+ |g'_n(0) - 1| + \log \left| \frac{1}{g''_n(0)} \right|$$

for all large n. The proof is completed as in Case 1, and the normality of $\widetilde{\mathcal{F}}$ is verified.

There is an interesting characterization of those families which, while satisfying (4.81), are nevertheless not normal, at least for analytic functions.

Theorem 4.5.20 *Let $\mathcal{F}$ be a family of analytic functions in U satisfying (4.81). If $\mathcal{F}$ is not normal at the origin, then there is a sequence $\{f_n\} \subseteq \mathcal{F}$ with*

$$\frac{f'_n}{f_n} \to \infty,$$

normally in a neighbourhood of the origin.

Proof. If $\mathcal{F}$ is not normal at the origin, there exists $\{f_n\} \subseteq \mathcal{F}$, $z_n \to 0$, $\rho_n \to 0$, and a nonconstant entire function f satisfying

$$f_n(z_n + \rho_n z) \to f(z),$$

uniformly on compact subsets of $\mathbb{C}$. By Theorem 4.5.13, there is a subsequence, again denoted by $\{f_n\}$ such that f'_n/f_n converges normally in U to a function ϕ that is either analytic or $\equiv \infty$. As f_n satisfies (4.81), f must be zero-free, and for the ℓ-th derivatives

$$\rho_n^\ell f_n^{(\ell)}(z_n + \rho_n z) = f_n(z_n + \rho_n z)^{(\ell)} \to f^{(\ell)}(z),$$

normally in $\mathbb{C}$. Thus, the derivatives of f are zero-free, and so

$$f(z) = e^{az+b}, \qquad a \neq 0.$$

The conclusion follows from the observation

$$\rho_n \frac{f'_n}{f_n}(z_n + \rho_n z) = \frac{f_n(z_n + \rho_n z)'}{f_n(z_n + \rho_n z)} \to a \neq 0.$$

That the same characterization cannot be applied to families of meromorphic functions is illustrated by the example $\left\{\frac{1}{nz}\right\}$.

In a similar vein, one may consider the family of analytic functions

$$\mathcal{F}_k = \{f : f(z)f^{(k)}(z) \neq 0, \ z \in \Omega\},$$

for some $k \in \mathbb{N}$. If $\Omega = \mathbb{C}$, that is, $f \in \mathcal{F}$ is entire, and $k \geq 2$, then $f(z) = e^{az+b}$, $a \neq 0$ (cf. Hayman [1964], Theorem 3.8). As the example $\{e^{nz}\}$ shows, such families cannot be expected to be normal. However, as in the preceding considerations, the associated family

$$\widetilde{\mathcal{F}}_k = \left\{\frac{f'}{f} : f \in \mathcal{F}_k\right\}$$

is normal (Schwick [1989]).

Finally we state a generalization of Theorem 4.4.19 also due to Schwick.

Theorem 4.5.21 *Let $\mathcal{F}$ be a family of meromorphic functions in a domain Ω such that for some $a, b \in \mathbb{C}$, $a \neq 0$, and analytic functions $a_1(z), \ldots, a_{k-1}(z)$ in Ω with $n \geq k + 4$,*

$$\sum_{m=1}^{k-1} a_m(z) f^{(m)}(z) + f^{(k)}(z) - af^n(z) \neq b, \qquad z \in \Omega.$$

Then $\mathcal{F}$ is normal.

We remark that Yang [1983] has proved this result under the additional assumption that the zeros of $f - c$, $c \neq 0$, $f \in \mathcal{F}$, are all of multiplicity greater than k.

5

General Applications

From its very inception, the theory of normal families has met with a wide range of applications. We have already encountered how it can be employed to prove such theorems as those of Picard, Schottky, Vitali-Porter, Lindelöf, and Julia. These were amongst its first uses, but there soon developed other fundamental applications to which we address ourselves here.

A very fruitful avenue has been in the application to various extremal problems, which amounts to finding a particular member of a family at which an extreme value is attained by a given continuous functional defined on the family. This was exactly the situation encountered in our proof of the Riemann Mapping Theorem. Another example arises with regard to the former Bieberbach Conjecture. The extremal method also lies at the heart of Hilbert's celebrated proof of the Dirichlet Principle.

A further significant application of the theory of normal families was made just after World War I by Fatou and Julia in their deep study of complex dynamical systems. The very notions of Fatou and Julia sets, as well as their attendant theory, rely heavily on the normal family milieu. Because of its singlular importance, we discuss the theory of complex dynamical systems in some detail.

A more recent application has been in the theory of normal functions, developed in order to study the boundary behaviour of meromorphic functions. Here again the definitions depend on the concept of a normal family and are related to the notion of invariant normal families.

We next consider families of harmonic functions which, it turns out, share various similar normality criteria with families of analytic functions. However, it is, in general, not the case that if $\{f\}$ is a normal family of analytic functions, then $\{\mathcal{R}e\, f\}$ is likewise normal. One important application of the harmonic theory is in the construction of the Martin kernel in the theory of positive harmonic functions. Harmonic analogues of the theorems of Schottky, Landau, Julia, and Montel are also demonstrated.

Another very classical field where normal families have played an important role is in the study of discrete/discontinuous groups of linear transformations. Here the special nature of the transformations engenders a sharper form of the FNT, which in turn can be used to discern when a discrete group is discontinuous.

5.1   Extremal Problems

Given a family $\mathcal{F}$ of analytic functions on a domain Ω, let $\phi : \mathcal{F} \to \mathbb{C}$ be a *complex functional*, that is, for each $f \in \mathcal{F}$, $\phi(f)$ is a complex number. A complex functional ϕ is *continuous* if $\phi(f_n) \to \phi(f)$ whenever $\{f_n\} \subseteq \mathcal{F}$ and f_n converges uniformly on compact subsets of Ω to $f \in \mathcal{F}$. For example, $\phi(f) = f(z_0)$, $z_0 \in \Omega$, is a continuous functional.

The extremal problem is initially to find some $f_0 \in \mathcal{F}$ for which

$$\sup_{f \in \mathcal{F}} |\phi(f)|$$

is attained at f_0. The problem is resolved by considering a compact family.

Theorem 5.1.1 *Let $\mathcal{F}$ be a compact normal family of analytic functions on a domain Ω, ϕ a continuous functional on $\mathcal{F}$. Then there exists a function $f_0 \in \mathcal{F}$ such that*

$$|\phi(f_0)| = \sup_{f \in \mathcal{F}} |\phi(f)|.$$

Proof. Suppose that

$$\sup_{f \in \mathcal{F}} |\phi(f)| = M \leq \infty.$$

Next let $\{f_n\}$ be a maximising sequence, that is, $\lim_{n \to \infty} |\phi(f_n)| = M$. As $\mathcal{F}$ is normal and compact, there exists a subsequence $\{f_{n_k}\}$ such that $f_{n_k} \to f_0 \in \mathcal{F}$ uniformly on compact subsets of Ω. Since ϕ is continuous, $|\phi(f_{n_k})| \to |\phi(f_0)|$, and consequently $|\phi(f_0)| = M (< \infty)$, as desired.

Similarly, we infer

$$\inf_{f \in \mathcal{F}} |\phi(f)| = |\phi(F_0)|$$

where $F_0 \in \mathcal{F}$. It is easy to find examples to show that the extremal function is not unique. Note also that the domain Ω may lie in $\widehat{\mathbb{C}}$.

Certain variational methods in the theory of *minimal surfaces* are based upon this approach (cf. Courant [1950], Chapter 3).

In Example 2.3, No. 5, it was shown that the schlicht class $\mathcal{S}$ was normal and compact in U. When $f \in \mathcal{S}$, $f(z) = z + a_2 z^2 + \ldots$, we define $\psi_n(f) = a_n$, for $n > 1$. Then each ψ_n is a continuous functional on $\mathcal{S}$. Indeed, suppose that $f_k \in \mathcal{S}$, $f_k \to f$, normally in U, with

$$f_k(z) = z + a_2^{(k)} z^2 + \ldots, \qquad f(z) = z + a_2 z^2 + \ldots .$$

If r, n are fixed, $C_r : |z| = r < 1$ and $n > 1$, then

$$\lim_{k \to \infty} \psi_n(f_k) = \lim_{k \to \infty} a_n^{(k)} = \lim_{k \to \infty} \frac{1}{2\pi i} \int_{C_r} \frac{f_k(\zeta) d\zeta}{\zeta^{n+1}}$$

$$= \frac{1}{2\pi i} \int_{C_r} \frac{f(\zeta) d\zeta}{\zeta^{n+1}} = a_n = \psi_n(f),$$

as desired. By the proceding theorem there exists $f_n \in \mathcal{S}$ such that $f_n(z)$ has the max $|a_n|$. That the *Koebe function*

$$k(z) = \frac{z}{(1-z)^2}$$

is the requisite extremal function *for each* n was known as the Bieberbach Conjecture, and was recently proved by L. de Branges [1985].

Essentially the same reasoning is applicable to

Example 5.1.2 Define

$$\mathcal{F} = \{f \text{ analytic in } \Omega : |f(z)| \leq 1\}.$$

Then, given z_0 in Ω, there exists f_0 in $\mathcal{F}$ such that $|f'(z_0)|$ is maximal in $\mathcal{F}$. This fact was made use of in our proof of the Riemann Mapping Theorem.

The extremal method of Theorem 5.1.1 has also found significant application in problems of conformally mapping a multiply connected domain onto one of several *canonical* domains. To illustrate the basic approach we shall treat one particular case: determining the conformal mapping of a multiply connected domain Ω onto a *parallel slit domain*. The latter consist of a subset of the extended complex plane, the complementary components of which are rectilinear slits making an angle θ with respect to the x-axis, or points. To take a simple example, note that the function

$$f_\theta(z) = z + \frac{e^{2i\theta}}{z}$$

is a univalent mapping of $|z| > 1$ onto the extended complex plane with a rectilinear slit from $-2e^{i\theta}$ to $2e^{i\theta}$. Here, and in what follows, we use the term *univalent* to mean one-to-one analytic, except for a possible (simple) pole which may be at infinity.

Given a multiply connected domain Ω in $\widehat{\mathbb{C}}$, consider the family

$$\mathcal{S}_\infty(\Omega) = \{f(z) \text{ univalent in } \Omega : f(z) \text{ has the Laurent expansion}$$
$$f(z) = z + \frac{a_1}{z} + \frac{a_2}{z^2} + \dots \text{ in a neighbourhood of } z = \infty\}.$$

We intend to demonstrate that there is an extremal function in $\mathcal{S}_\infty(\Omega)$ which attains the

$$\max_{f \in \mathcal{S}_\infty(\Omega)} \mathcal{R}e\,(e^{-2i\theta}a_1),$$

and that this function is the desired conformal mapping. This line of reasoning dates back to de Possel [1931]; cf. also Grötzsch [1932].

Firstly, the following growth estimate is required.

Lemma 5.1.3 *If*
$$f(z) = z + a_0 + \frac{a_1}{z} + \dots$$
is univalent in $|z| > R$, *then*
$$|f(z) - a_0| \leq 2|z|.$$

Proof. Given $|z_0| > R$, define
$$w = F(z) = \frac{1}{z_0} f(z_0 z) - \frac{a_0}{z_0} = z + \frac{a_1}{z_0^2 z} + \dots,$$
which is univalent in $|z| > 1$. Then consider the function
$$G(z) = \frac{1}{F\left(\frac{1}{z}\right) - c} = z + cz^2 + \dots$$

for any c not in the range of F. Clearly G belongs to the schlicht class S (cf. §2.3) and as a consequence its second Taylor coefficient satisfies $|c| \leq 2$. Therefore, the entire boundary of $F(\{|z| > 1\})$ is contained in $|w| \leq 2$. Specifically, $|F(1)| \leq 2$ which gives the result
$$|f(z) - a_0| \leq 2|z|, \qquad |z| > R.$$

Our main result also depends upon the following observation.

Lemma 5.1.4 *In the family*
$$\mathcal{S}_R = \left\{ f(z) \text{ univalent in } |z| > R : f(z) = z + \frac{a_1}{z} + \dots \right\},$$
given $\theta \in \mathbb{R}$, *the function* $f_\theta(z) = z + \frac{R^2 e^{2i\theta}}{z}$, *which maps* $|z| > R$ *onto the plane with a rectilinear slit at angle* θ, *uniquely solves the extremal problem:*
$$\max_{f \in \mathcal{S}_R} \mathcal{Re}\left(e^{-2i\theta} a_1\right),$$
giving the maximum value R^2.

Proof. Normalizing the functions of $\mathcal{S}_R$ by
$$F(z) = \frac{1}{R} f(Rz) = z + \frac{a_1}{R^2 z} + \dots$$

yields the functions $\{F(z)\}$ which are univalent in $|z| > 1$. At this stage we wish to invoke the Area Theorem (cf., e.g., Hayman [1958], p. 3, or Duren [1983], p. 29) which asserts that any function $g(\zeta) = \zeta + \frac{a_1}{\zeta} + \dots$ which is univalent in $|\zeta| > 1$ satisfies the condition
$$\sum_{n=1}^{\infty} n|\alpha_n|^2 \leq 1.$$

In particular, $|\alpha_1| \leq 1$ with equality holding if and only if $\alpha_2 = \alpha_3 = \ldots = 0$, i.e., for the function $g(\zeta) = \zeta + \frac{e^{i\phi}}{\zeta}$, discussed above. Applying this result to the present setting gives $|a_1| \leq R^2$, with equality holding if and only if $f(z) = z + \frac{R^2 e^{i\phi}}{z}$. Hence

$$\mathcal{R}e\left(e^{-2i\theta}a_1\right) \leq R^2,$$

with equality holding if and only if $f(z) = z + \frac{R^2 e^{2i\theta}}{z}$.

Theorem 5.1.5 *Every multiply connected domain Ω in $\widehat{\mathbb{C}}$ can be mapped by a univalent function onto a parallel slit domain, with slits of presecribed angle θ. Moreover, if ζ is a given point of Ω which is mapped to ∞, then the expansion of the mapping about $z = \zeta$ has the form*

$$w = \frac{1}{z - \zeta} + a_1(z - \zeta) + \ldots \quad or \quad w = z + \frac{a_1}{z} + \ldots$$

according to whether ζ is finite or ∞.

Proof. Consider first the case $\zeta = \infty$. Then all of $\partial\Omega$ is contained in $|z| < R$ for R sufficiently large. Thus every f in $\mathcal{S}_\infty(\Omega)$ is univalent in $|z| > R$ and has the expansion $f(z) = z + \frac{a_1}{z} + \ldots$ there. In view of Lemma 5.1.3, $\mathcal{S}_\infty(\Omega)$ is locally bounded in $R < |z| < \infty$, and if $|z| = R' > R$, then $|f(z)| \leq 2R'$. The latter implies that $|f(z)| \leq 2R'$ for all $z \in \Omega \cap \{|z| < R'\}$ due to the univalence of $f \in \mathcal{S}_\infty(\Omega)$. To see this, observe that if $|w| > 2R'$, then it follows from Rouché's theorem that the equations $f(z) = w$ and $f(z) = \infty$ have equally many roots in $|z| > R'$, i.e., exactly one. Since $f(z)$ is univalent in Ω, the equation $f(z) = w$ ($|w| > 2R'$) will have no other roots in Ω, and thus none in $\Omega \cap \{|z| < R'\}$, so that $|f(z)| \leq 2R'$ there. We conclude that $\mathcal{S}_\infty(\Omega)$ is normal in $\Omega - \{\infty\}$. Moreover, if $f_j \to f$ normally in $\Omega - \{\infty\}$, and $f_j(z) = z + \sum_{n=1}^{\infty} \frac{a_n^{(j)}}{z^n}$, $j = 1, 2, 3, \ldots$ ($R < |z| < \infty$), then $a_n^{(j)} \to a_n$ and f is analytic univalent, with $f(z) = z + \sum_{n=1}^{\infty} \frac{a_n}{z^n}$ ($R < |z| < \infty$), that is, $\mathcal{S}_\infty(\Omega)$ is compact in $\Omega - \{\infty\}$. Note also that the family of analytic univalent functions $f(z)$ defined on $\Omega - \{\infty\}$ with the expansion $f(z) = z + \frac{a_1}{z} + \ldots$ in a neighbourhood of ∞ coincides with $\mathcal{S}_\infty(\Omega)$.

Next observe that the functional

$$\phi_\theta(f) = \mathcal{R}e\left(e^{-2i\theta}a_1\right), \qquad f \in \mathcal{S}_\infty(\Omega),$$

is continuous. An appeal to Theorem 5.1.1 yields the existence of $F(z) = z + \frac{A_1}{z} + \ldots \in \mathcal{S}_\infty(\Omega)$ such that

$$\max_{f \in \mathcal{S}_\infty(\Omega)} \mathcal{R}e\left(e^{-2i\theta}a_1\right) = \mathcal{R}e\left(e^{-2i\theta}A_1\right).$$

It remains to show that the extremal function $F(z)$ is the desired conformal mapping. Suppose that the domain $\Omega^* = F(\Omega)$ has at least one boundary component β that is a continuum but not a rectilinear slit of inclination θ. Let Ω_1^* be the unbounded component of the complement of β. Then $\Omega^* \subseteq \Omega_1^*$ since Ω^* is connected, unbounded, and lies in the complement of β. Moreover, Ω_1^* is simply connected since its complement with respect to $\widehat{\mathbb{C}}$ is connected. Now let $w' = g(w)$ be a univalent mapping of Ω_1^* onto the w'-plane with a rectilinear slit of inclination θ, having the expansion

$$g(w) = w + \frac{a_\theta}{w} + \dots$$

in a neighbourhood of $w = \infty$ (cf. Remarks 2.6.1 subsequent to the Riemann Mapping Theorem). Then the composition

$$F^*(z) = g\bigl(F(z)\bigr)$$

maps the domain Ω univalently and has the expansion

$$F^*(z) = z + \frac{A_1 + a_\theta}{z} + \dots$$

near $z = \infty$, so that $F^* \in \mathcal{S}_\infty(\Omega)$. Therefore

$$\begin{aligned}
\mathcal{R}e\left(e^{-2i\theta}(A_1 + a_\theta)\right) &= \mathcal{R}e\left(e^{-2i\theta}A_1\right) + \mathcal{R}e\left(e^{-2i\theta}a_\theta\right) \\
&\leq \mathcal{R}e\left(e^{-2i\theta}A_1\right),
\end{aligned}$$

giving $\mathcal{R}e\left(e^{-2i\theta}a_\theta\right) \leq 0$.

On the other hand, in view of the Remarks 2.6.1, a domain $|W| > R'$ can be mapped univalently onto Ω_1^* by

$$h(W) = W + b_0 + \frac{b_1}{W} + \dots, \qquad |W| > R'.$$

Then the function

$$w' = g\bigl(h(W)\bigr) - b_0 = W + \frac{b_1 + a_\theta}{W} + \dots$$

maps $|W| > R'$ univalently onto the w'-plane with rectilinear slit at angle θ, whereas the mapping $h(W) - b_0$ maps $|W| > R'$ onto some other domain. In accordance with Lemma 5.1.4

$$\mathcal{R}e\left(e^{-2i\theta}b_1\right) < \mathcal{R}e\left(e^{-2i\theta}b_1 + e^{-2i\theta}a_\theta\right),$$

implying $\mathcal{R}e\left(e^{-2i\theta}a_\theta\right) > 0$, contradicting the result of the preceding paragraph. This proves the first part of the theorem.

In the case of finite ζ, an application of the mapping $\frac{1}{z-\zeta} = z'$ transforms Ω onto a domain Ω' which contains the point $z' = \infty$, reducing the problem to the previous case. The proof of the theorem is herewith complete.

From the proof we make note of the fact that the class

$$S_1 = \left\{ g(z) \text{ univalent in } |z| > 1 : g(z) = z + \frac{b_1}{z} + \frac{b_2}{z^2} + \dots \right\}$$

is normal and compact.

For the conformal mappings of a multiply connected domain onto other canonical regions, refer to Nehari [1952], Chapter 7, Goluzin [1969], Chapter 5, Duren [1983], Chapter 10.

5.2   Dynamical Systems

An important application of the theory of normal families is the study of iterations of a function, giving rise to discrete dynamical systems. Early work in this field was initiated by Julia [1918] and Fatou [1919, 1920], who exploited the newly developed theory of Montel. For a recent survey article, see Blanchard [1984], or for further details, the books by Peitgen and Richter [1986], Devaney [1987], and Beardon [1991].

We investigate the nature of a discrete dynamical system generated by a meromorphic mapping on the Riemann sphere. As such maps are rational functions on $\widehat{\mathbb{C}}$ (cf. Palka [1991], Theorem 4.3), consider

$$R(z) = \frac{P(z)}{Q(z)} : \widehat{\mathbb{C}} \to \widehat{\mathbb{C}}, \qquad P(z), Q(z) \not\equiv 0,$$

where $P(z)$ and $Q(z)$ are polynomials with complex coefficients having no common factors. For the theory, it is assumed that the *degree* of $R(z)$,

$$\deg(R) = \max\{\deg(P), \deg(Q)\} \geq 2,$$

in order to avoid certain trivialities. The degree can be viewed as the number of roots in $\widehat{\mathbb{C}}$ of the equation $R(z) - a = 0$, where a is an arbitrary constant, and the roots are counted according to multiplicity. That is, $R(z)$ is a *d-fold* mapping of $\widehat{\mathbb{C}}$ onto itself. A rational map $w = R(z)$ considered on the Riemann sphere is also an *analytic* map as the value $w = \infty$ loses its special significance.

A simple equivalence relation we shall have occasion to use it that of *conjugation*. Suppose $R(z)$ and $S(z)$ are two rational functions such that the diagram

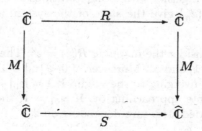

commutes, i.e., $S = M \circ R \circ M^{-1}$, where M is a Möbius transformation. Then we say that R and S are analytically *conjugate*. Clearly, if R and S are conjugate by M, then R^n and S^n (the iterates of R and S, cf. below) are conjugate by M as well. This means that R and S generate "equivalent" dynamical systems.

For example, it is readily confirmed that an arbitrary quadratic polynomial

$$R(z) = az^2 + 2bz + d$$

is conjugate to the quadratic polynomial $p(z) = z^2 + c$, where $c = ad + b - b^2$, via the Möbius function $M(z) = az + b$.

Thus, in order to investigate the dynamics of quadratic polynomials, it suffices to consider only those of the form $p(z) = z^2 + c$.

Periodic Points. For $z_0 \in \mathbb{C}$, define successive *iterates* of z_0 by the recursive relation $z_0 = R^0(z_0)$, and

$$
\begin{aligned}
z_{n+1} &= R(z_n) \\
&= R\big(R^n(z_0)\big), \qquad n = 0, 1, 2, \ldots,
\end{aligned}
$$

where $R^n = R \circ R \circ \ldots \circ R$, n times. These iterates comprise the *forward orbit*, $Or^+(z_0)$, of z_0. There is also the *backward orbit*

$$Or^-(z_0) = \{ z \in \widehat{\mathbb{C}} : R^m(z) = z_0 \text{ for some } m = 0, 1, 2, \ldots \}.$$

Denote by $z_{-m} = R^{-m}(z_0)$, an antecedent in $Or^-(z_0)$ satisfying $R^m(z_{-m}) = z_0$. If for some $z_n \in Or^+(z_0)$, $R^n(z_0) = z_n = z_0$ and $z_m \neq z_0$ for $0 < m < n$, then z_0 is a *periodic point of period n*, or *fixed point of order n*. If $n = 1$, z_0 is simply called a *fixed point*. If z_0 is a periodic point of period n, then $Or^+(z_0)$ consists of n points

$$\gamma = \{ z_0, z_1 = R(z_0), \ldots, z_{n-1} = R^{n-1}(z_0) \}$$

and is called a *periodic* orbit, or *n-cycle*. Note that each $z_i \in \gamma$ is a fixed point of the mapping $R^n(z)$. If z_0 itself is not periodic, but some iterate $R^m(z_0)$ is periodic, then z_0 is called *preperiodic*.

In the present context, a *dynamical system* consists of a rational function $R(z)$ (the *dynamic*), together with all its iterations as z ranges over $\widehat{\mathbb{C}}$. The iterations $z_{n+1} = R(z_n)$ give the *state* of the system as a function of its previous state.

Example 5.2.1 Consider the mapping $R(z) = z^2$. Then $R(z)$ has exactly three fixed points, $0, 1$, and ∞. Moreover, if $0 < |z_0| < 1$, then $Or^+(z_0)$ is a sequence of points converging to the origin. If $1 < |z_0| < \infty$, then $Or^+(z_0)$ is a sequence of points approaching ∞. If $|z_0| = 1$, then $Or^+(z_0)$ lies on $|z| = 1$ and has a varied structure depending on z_0.

Let z_0 be a fixed point of order n. The stability of z_0 is characterized by considering the *eigenvalue* of z_0, namely, the derivative $\lambda = \lambda_{z_0} = (R^n)'(z_0)$ for $z_0 \in \mathbb{C}$, (or $\frac{1}{(R^n)'(z_0)}$ if $z_0 = \infty$). By the chain rule, for $z_0 \in \mathbb{C}$, $\lambda = R'(z_0) \cdot R'(z_1) \cdots R'(z_{n-1})$, and so λ has the same value at each point of the periodic orbit. Thus λ is an *invariant* of the orbit γ. Similiarly for $z_0 = \infty$. Note that λ is also invariant under conjugation. If

(i) $0 < |\lambda| < 1$, then z_0 is an *attractive* fixed point;

(ii) $\lambda = 0$, then z_0 is a *superattractive* fixed point;

(iii) $|\lambda| > 1$, then z_0 is a *repulsive (repelling)* fixed point;

(iv) $|\lambda| = 1$, then z_0 is an *indifferent (neutral)* fixed point;

The nature of the fixed points plays a vital role in describing the behaviour of the dynamical system.

Julia and Fatou Sets. At this juncture we require certain aspects of the theory of normal families developed in Chapters 2 and 3.

Define the *Julia* set as

$$J_R = \big\{ \zeta \in \widehat{\mathbb{C}} : \{R^n(z)\} \text{ is } not \text{ normal at } \zeta \big\}.$$

Historically this set has been denoted by F. However, in modern parlance, we denote the complement by $F_R = \widehat{\mathbb{C}} - J_R$, and call F_R the *Fatou (or stable) set*. Hence, about each point $\zeta \in F_R$, there is a neighbourhood N_ζ in which $\{R^n(z) \mid N_\zeta\}$ is a normal family. Therefore, F_R is an *open* set, the connected components of which are the maximal domains of normality of $\{R^n(z)\}$, and J_R is a *closed* set.

In our former example, $R(z) = z^2$, if $|z_0| < 1$, then in any disk $D = D(z_0; r) \subseteq U$, $\{R^n(z) \mid D\}$ converges normally to the identically zero function. Thus $\{|z| < 1\} \subseteq F_R$, and similarly, $\{|z| > 1\} \subseteq F_R$ where here the limit function is identically infinite. Clearly for $|z_0| = 1$, there is no disk about z_0 in which $\{R^n(z)\}$ forms a normal family, so that $\{|z| = 1\} = J_R$.

Theorem 5.2.2 (i) *If $Or^+(z_0)$ is a (super) attractive periodic orbit, then $Or^+(z_0) \subset F_R$.*

(ii) *If $Or^+(z_0)$ is a repulsive periodic orbit, then $Or^+(z_0) \subseteq J_R$.*

Proof. (i) We may suppose that z_0 is an attractive fixed point, and $z_0 \neq \infty$. Then, in a sufficiently small disk $D = D(z_0; r)$,

$$\left| \frac{R(z) - R(z_0)}{z - z_0} \right| < \sigma < 1,$$

that is, $|R(z) - z_0| < \sigma |z - z_0|$. Whence $|R^n(z) - z_0| < \sigma^n r$, $n = 1, 2, 3, \ldots$, implying $\{R^n(z)\}$ converges normally to z_0 in D, and $z_0 \in F_R$; indeed $D \subseteq F_R$.

(ii) is deduced similarly.

Thus for an attractive fixed point $z_0 \in \widehat{\mathbb{C}}$, we are led to define the set

$$A(z_0) = \{z \in \widehat{\mathbb{C}} : R^n(z) \to z_0 \text{ as } n \to \infty\},$$

called the *basin (domain) attraction of* z_0. In general, it consists of a countable number of (open) domains and contains all points z whose forward orbits, $Or^+(z)$, approach z_0, including all the points of $Or^-(z_0)$. In Example 5.2.1 the basins of attraction are $\{|z| < 1\}$ and $\{|z| > 1\}$. The *immediate basin of attraction* $A^*(z_0)$ is the connected component of $A(z_0)$ containing z_0. For a periodic orbit of period n,

$$\gamma = \{z_0, z_1 = R(z_0), \ldots, z_{n-1} = R^{n-1}(z_0)\},$$

then the *immediate basin of attraction of* γ is given by

$$A^*(\gamma) = \bigcup_{i=0}^{n-1} A^*(z_i, R^n),$$

where $A^*(z_i, R^n)$ is the immediate attractive set for the attractive fixed point z_i of the mapping R^n.

Exceptional Points. An immediate consequence of the FNT on $\widehat{\mathbb{C}}$ (cf. §3.2 and the comments following Theorem 3.3.2) is

Proposition 5.2.3 *Given* $\zeta \in J_R$, *if* Δ *is an arbitrary neighbourhood of* ζ, *then the set*

$$E_\Delta = \widehat{\mathbb{C}} - \bigcup_{n>0} R^n(\Delta)$$

contains at most two points, called exceptional points.

For a polynomial $p(z)$, note that $z = \infty$ is a superattractive fixed point and thus $\infty \in F_p$, $J_p \subseteq \mathbb{C}$. Moreover, $\cup_{n>0} p^n(\mathbb{C}) = \mathbb{C}$ implies ∞ is an exceptional point for $p(z)$.

For each $\zeta \in J_R$, define

$$E_\zeta = \bigcup E_\Delta$$

by taking the union over all neighbourhoods of ζ. It is obvious that $0 \leq \overline{\overline{E}}_\zeta \leq 2$, and that E_ζ is independent of Δ for all Δ sufficiently small. Thus

$$\widehat{\mathbb{C}} - E_\zeta$$
$$= \{z : \text{ for each } \varepsilon > 0 \text{ there exists } z_1 \text{ with } |\zeta - z_1| < \varepsilon \text{ and } z \in Or^+(z_1)\}.$$

The set of exceptional points in $\widehat{\mathbb{C}}$ characterizes the mapping $R(z)$ in a fundamental way.

Theorem 5.2.4 *Let $\zeta \in J_R$, and suppose that $E_\zeta \neq \emptyset$.*

(i) *If $\overline{\overline{E}}_\zeta = 1$, then $R(z)$ is conjugate to a polynomial.*

(ii) *If $\overline{\overline{E}}_\zeta = 2$, then $R(z)$ is conjugate to the mapping $z \rightarrow z^{\pm d}$, where $d = \deg(R)$.*

Proof. Since, by the definition of E_ζ, $R^{-1}(E_\zeta) \subseteq E_\zeta$, the surjectivity of R implies that $R^{-1}(E_\zeta) = E_\zeta$, so that E_ζ consists of either a single fixed point, two fixed points, or a 2-cycle, as it is assumed to be nonempty.

(i) Suppose $E_\zeta = \{\alpha\}$. Choose a Möbius transformation $M(z)$ such that $M(\alpha) = \infty$. Then the rational mapping

$$p(z) = M \circ R \circ M^{-1}(z)$$

has no poles in $\mathbb{C}$ as only $p(\infty) = \infty$, whence $p(z)$ is a polynomial.

(ii) Suppose $E_\zeta = \{\alpha, \beta\}$, and choose a Möbius transformation $M(z)$ satisfying $M(\alpha) = \infty$, $M(\beta) = 0$. Again set $p(z) = M \circ R \circ M^{-1}(z)$ and consider two cases:

(a) Both 0 and ∞ are fixed points of $p(z)$. As in (i), $p(z)$ is a polynomial. Since $p(0) = 0$ and no other point can map to 0, the origin is a zero of multiplicity d. It follows that $p(z) = Kz^d$, and by considering a conjugation with either an expansion or contraction, the constant K may be set equal to 1.

(b) If the points 0 and ∞ form a 2-cycle, then similarly we arrive at $p(z) = z^{-d}$.

Corollary 5.2.5 (a) $E_\zeta \subseteq F_R$; (b) E_ζ *is independent of the choice of* $\zeta \in J_R$.

Proof. (a) In Case (i) of the theorem, E_ζ consists of a superattractive fixed point; in Case (ii), E_ζ consists of two superattractive fixed points. The conclusion then follows from Theorem 5.2.2.

(b) For any $\zeta \in J_R$, if $E_\zeta \neq \emptyset$, then R is conjugate to either a polynomial or to the mapping $z \rightarrow z^{\pm d}$. Taking the domain $D = \mathbb{C}$ in (i) and $D = \widehat{\mathbb{C}} - \{0, \infty\}$ in (ii) gives $E_D = E_\zeta$, and E_ζ is independent of $\zeta \in J_R$. (Here we have blurred the distinction between R and its conjugate).

As a consequence of (b) we may denote the set of exceptional points by E_R.

Characterizations of J_R. We are now able to state a number of properties which characterize the Julia set J_R.

Theorem 5.2.6 (i) $R(J_R) = J_R = R^{-1}(J_R)$.

(ii) J_R *is a non-empty perfect set, that is, J_R is equal to its set of accumulation points.*

(iii) *If J_R contains an* <u>*interior point*</u>*, then $J_R = \widehat{\mathbb{C}}$.*

(iv) *For $z \in J_R$, $J_R = \overline{\bigcup_{n \geq 0} R^{-n}(z)}$.*

(v) *If α is an attractive fixed point of R, then $A(\alpha) \subseteq F_R$ and $\partial A(\alpha) = J_R$.*

(vi) *The set of repulsive fixed points of all orders is dense in J_R.*

Proof. (i) Let $\zeta \in J_R$ and $\zeta_1 = R(\zeta) \in R(J_R)$. If $\{R^n\}$ converges normally in a neighbourhood of ζ_1, it would follow that $\{R^{n+1}\}$ converges normally in a neighbourhood of ζ. As this would be a contradiction, $R(J_R) \subseteq J_R$. Furthermore, if $\{R^n\}$ converges normally in a neighbourhood of ζ_{-1}, then $\{R^{n-1}\}$ converges likewise in a neighbourhood of ζ, i.e., $R^{-1}(J_R) \subseteq J_R$. Applying R to this latter inclusion and R^{-1} to the former, and noting that $J_R \subseteq R^{-1}(R(J_R))$, proves (i).

The two equalities thus established are referred to respectively as the *forward* and the *backward invariance* of J_R. Both together are called the *complete invariance* of J_R. Note that F_R has the same property.

(ii) First suppose that $J_R = \emptyset$, so that $\{R^n\}$ is normal in $\widehat{\mathbb{C}}$. Then there is a subsequence $\{R^\nu\}$ which converges spherically uniformly in all of $\widehat{\mathbb{C}}$ to a limit function $S(z)$. Then $S(z)$ is a rational function of degree k; $S(z)$ may be a constant, either finite or infinite. Let a be an arbitrary finite value of $S(z)$ if $S(z)$ is not constant, and choose a to be distinct from $S(z)$ if the latter is constant. Then the equation $S(z) - a = 0$ has exactly k roots, and likewise for $R^\nu(z) - a = 0$ for all sufficiently large ν. But $\deg(R^\nu) \to \infty$ as $\nu \to \infty$, and this contradiction proves $J_R \neq \emptyset$.

In order to show J_R is a perfect set, we first claim: If $a \in J_R$, then there is some $b \in J_R$ such that $a \in Or^+(b)$, yet $b \notin Or^+(a)$. In fact, for a nonperiodic point a, it suffices to choose $b \in R^{-1}(a)$. On the other hand, suppose that $a \in J_R$ is periodic with period n. Denote $S = R^n$ and consider the equation

$$S(z) = a. \tag{5.1}$$

If $S^{-1}(a) = \{a\}$, then choose a Möbius transformation M with $M(a) = \infty$. Hence the conjugate

$$p(z) = M \circ S \circ M^{-1}(z)$$

must be a polynomial as $p(z)$ has no poles in $\mathbb{C}$. Since $z_0 = \infty$ is always a superattractive fixed point for a polynomial, $\infty \in F_p$, that is, $a \in F_S$, contradicting $a \in J_R$. We deduce that there exists another point $b \in S^{-1}(a)$, and $a \in Or^+(b)$. However, as a is the only solution of (5.1) belonging to $Or^+(a)$, we have $b \notin Or^+(a)$, and the assertion is proved.

Returning to J_R, we prove that every point $a \in J_R$ is an accumulation point of J_R. Take an arbitrary neighbourhood Δ of a, and consider $b \in J_R$ with $b \notin Or^+(a)$, $a \in Or^+(b)$ as above. Then $b \notin E_R$ as $J_R \cap E_R = \emptyset$, and so $b \in R^k(\Delta)$ for some $k \in \mathbb{N}$. Choosing $c \in \Delta$ with $R^k(c) = b$, we observe that $c \neq a$ since $b \notin Or^+(a)$. But $c \in J_R$ by the backward invariance of J_R, making a an accumulation point, and therefore J_R is perfect.

(iii) Suppose J_R contains an interior point a and take an open disk Δ about a with $\Delta \subseteq J_R$. Then $\{R^n(z)\}$ is not normal in Δ, and hence $\cup_{n>0} R^n(\Delta)$ omits at most two points of $\widehat{\mathbb{C}}$. Since $\cup_{n>0} R^n(\Delta) \subseteq J_R$ by forward invariance, and J_R is closed, the conclusion follows.

(iv) Assume $z \in J_R$ and take a neighbourhood Δ of any $w \in J_R$. Since $z \notin E_R$, we have $z = R^k(\zeta)$ for some $k \in \mathbb{N}$, $\zeta \in \Delta$. Thus $\zeta \in \cup_{n \geq 0} R^{-n}(z)$, and since Δ is arbitrary, w is an accumulation point of $\cup_{n \geq} R^{-n}(z)$, i.e., $J_R \subseteq \overline{\cup_{n \geq 0} R^{-n}(z)}$. Furthermore, $\cup_{n \geq 0} R^{-n}(z) \subseteq J_R$ by the backward invariance of J_R, and since J_R is closed, $\overline{\cup_{n \geq 0} R^{-n}(z)} = J_R$, as desired.

(v) Take $z \in A(\alpha)$. As $\alpha \in F_R$ according to Theorem 5.2.2(i) and F_R is open, consider a disk $D(\alpha; r) \subseteq F_R$. Then $R^n(z) \in D(\alpha; r)$ for all n sufficiently large, whence $z \in F_R$ by the backward invariance of F_R, and $A(\alpha) \subseteq F_R$. Finally, for $z \in J_R$, any neighbourhood N of z satisfies $N \cap A(\alpha)' \neq \emptyset$ since $z \in A(\alpha)'$. Moreover, as $\{R^n(N)\}$ omits at most two points, iterations of some points of N must lie in $A(\alpha)$, implying $N \cap A(\alpha) \neq \emptyset$, and so $J_R \subseteq \partial A(\alpha)$. On the other hand, if $z \in \partial A(\alpha)$, any neighbourhood N of z contains points of $A(\alpha)$ and $A(\alpha)'$, so that $\{R^n\}$ cannot be normal in N and therefore $z \in J_R$; whence $J_R = \partial A(\alpha)$.

(vi) The proof given here follows the classical approach. However, there is a more direct proof, based on the Ahlfors Five Islands Theorem along the lines of Baker's proof for the entire case (see Theorem 5.2.11).

From Theorem 5.2.2 we know that every repulsive periodic point belongs to J_R, and as J_R is closed, the closure of the set of repulsive fixed points is contained in J_R. In order to prove equality, we must make a brief diversion to establish some preliminary results.

Definition 5.2.7 *Given a number $v \in \widehat{\mathbb{C}}$, a point $c \in \widehat{\mathbb{C}}$ is a* **critical point** *(with respect to mapping R) if it is a solution of $R(z) - v = 0$ having multiplicity greater than 1. The number v is then a* **critical value**.

We shall denote by C_R the set of critical points of R and by V_R the set of critical values. It is evident that $R(C_R) = V_R$. Moreover, for finite z, if $R'(z) \neq 0, \infty$, then R is injective in a neighbourhood of z, implying $\overline{C_R} < \infty$. Indeed, the inequality $\overline{C_R} \leq 2(d-1)$ follows from the Riemann-Hurwitz Theorem, where $d = \deg(R)$ (cf. Blanchard [1984]).

If $w = R(z)$ is not a critical value, then $R^{-1}(\{w\}) = \{z_1, z_2, \ldots, z_d\}$, say, where no z_i, is a critical point of R. In this case, R maps some neighbourhood N_i of z_i bijectively onto a neighbourhood N of w, $i = 1, 2, \ldots, d$, and hence $R \mid N_i$ has branches

$$R_i^{-1} : N \to N_i, \qquad i = 1, 2, \ldots, d,$$

of R^{-1} at w.

Thus we maintain

Theorem 5.2.8 $J_R \subseteq \overline{\{periodic\ points\}}$.

Proof. Define the subset K_R of J_R by

$$K_R = J_R - \{\text{critical values of } R^2\}.$$

As J_R is a perfect set and K_R differs from J_R by only a finite subset, it suffices to prove that $K_R \subseteq \{\text{periodic points}\}$. Let $w \in K_R$ and consider an open set V containing w. Since $d \geq 2$, $R^{-2}(\{w\})$ consists of at least four points, and denote by $\zeta_1, \zeta_2, \zeta_3$ three of these that are distinct from w. Next we find pairwise disjoint open neighbourhoods N_1, N_2, N_3, of $\zeta_1, \zeta_2, \zeta_3$, respectively, such that $R^2 : N_i \to V' \subseteq V$ is a homeomorphism, and set $S_i : V' \to N_i$ to be the inverse of $R^2 \mid N_i$, $i = 1, 2, 3$.

Now suppose that for all $z \in V'$

$$R^n(z) \neq S_i(z), \qquad n \geq 1, \ i = 1, 2, 3.$$

Then $\{R^n\}$ is normal in V' contradicting the fact that $w \in J_R$. As a consequence, there exists some $z \in V'$ and some $m \geq 1$ such that $R^m(z) = S_j(z)$ for some $j, 1 \leq j \leq 3$. Therefore,

$$R^{m+2}(z) = R^2 S_j(z) = z,$$

and so $z \in V'$ is a periodic point of R, and we conclude that $K_R \subseteq \overline{\{\text{periodic points}\}}$.

As this juncture we also need to know whether the number of attractive cycles of R is finite or not. To this end there is

Theorem 5.2.9 *The number of attractive cycles of R is at most $2(d-1)$, where $d = \deg R$.*

Proof. We claim that for each attractive cycle γ, the immediate basin of attraction $A^*(\gamma)$ contains at least one critical point, of which there are only $2(d-1)$ in number. Consider first the case where α is an attractive fixed point of order one, and suppose that $A^*(\alpha)$ contains no critical value. Choose a disk neighbourhood N of α with $N \subseteq A^*(\alpha)$ and take a branch R_*^{-1} of the inverse $R^{-1}(z)$ that satisfies $R_*^{-1}(\alpha) = \alpha$. Then R_*^{-1} can be extended to be meromorphic in N owing to the absence of critical values and the Monodromy Theorem (§1.4) and satisfies $R_*^{-1}(N) \subset A^*(\alpha)$. Similarly, we can define the functions

$$R_*^{-n}(z) = R_*^{-1}\big(R_*^{-(n-1)}(z)\big), \qquad n = 2, 3, \ldots, \ z \in N,$$

so that $R_*^{-n}(z) \subseteq A^*(\alpha)$. This results in a sequence of meromorphic functions $\{R_*^{-n}\}$ defined on N that omits at least three values (in particular,

J_R). The normality of $\{R_*^{-n}\}$, however, contradicts the fact that α is a repulsive fixed point of R_*^{-1}.

In the case where α is part of an attractive cycle $\{\alpha, \alpha_1, \ldots, \alpha_k\}$, then the functions $R_*^{-n}(z)$ are generated by

$$R_*^{-1}(\alpha) = \alpha_k, R_*^{-2}(\alpha) = a_{k-1}, \ldots, R_*^{-k}(\alpha) = \alpha,$$

and we proceed as above.

As an immediate consequence we have

Corollary 5.2.10 *The number of attractive periodic points is finite.*

Returning to the proof of (vi), we have thus far deduced

$$\overline{\{\text{repulsive periodic points}\}} \subseteq J_R \subseteq \overline{\{\text{periodic points}\}}.$$

The difference between the set of periodic points and the set of repulsive periodic points is the sets of attractive and indifferent periodic points. The former has finite cardinality as stated above. The proof that the set of indifferent periodic points is finite is quite technical and we refer the interested reader to Blanchard [1984] or Beardon [1991], §9.6, for the details. Assuming this fact, as J_R is a perfect set, we obtain (vi).

A few comments are in order. Property (iv) gives an expedient method for generating a dense subset of J_R. A repulsive fixed point is found and then its inverse orbit determined.

Moreover, most Julia sets are fractals, that is, a set whose Hausdorff dimension is nonintegral (cf. Mandelbrot [1982]). The structure of the Julia set is one of the most fascinating aspects of a complex dynamical system, and extremely intricate structures arise out of elementary polynomials.

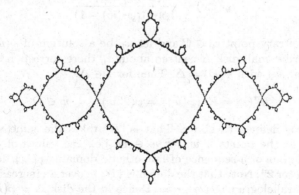

Figure 5.1 Julia set of $R(z) = z^2 - 1$

Although the Julia set depicted in Figure 5.1 is a connected set, in general this may or may not be the case. Indeed, for any polynomial $R(z) = z^2 - p$, if $-\frac{1}{4} \le p \le 2$, then J_R is connected, and if $p < -\frac{1}{4}$ (Figure 5.2) or $p > 2$, then J_R is *totally disconnected*, that is, the only connected subsets are singleton sets (cf. Brolin [1965]).

Figure 5.2 Julia set of $R(z) = z^2 + 0.3$

Regarding (iii), Lattès [1918] demonstrated that the rational function

$$R(z) = \frac{(z^2 + 1)^2}{4z(z^2 - 1)}$$

has $J_R = \widehat{\mathbb{C}}$.

To see this, let $\wp(w)$ be the Weierstrass $\wp$-function (§1.7), and observe that $\wp(2w)$ is a rational function of $\wp(w)$, namely

$$\wp(2w) = \frac{\big(\wp^2(w) + 1\big)^2}{4\wp(w)\big(\wp^2(w) - 1\big)}.$$

Take an arbitrary point $z_0 \in \mathbb{C}$ and let w_0 be a solution of $\wp(w) = z_0$. To any arbitrarily small disk Δ centred about z_0 there corresponds a domain D containing w_0 with $\wp(D) = \Delta$. Then for $z \in \Delta$,

$$R^n(z) = R^n\big(\wp(w)\big) = \wp(2^n w), \qquad w \in D,$$

when $R(z)$ is defined as above. That is, R^n takes the same values in Δ as $\wp$ takes at the points $2^n w$, for $w \in D$, i.e., the values of $\wp(w)$ for w belonging to one of a sequence of homothetic domains $\{D_n\}$ derived from D by the factor 2^n. Note that the domains $\{D_n\}$ cover an increasing number of period parallelograms as $n \to \infty$. Hence in the disk $\Delta = \wp(D)$, R^n will take each value ζ a certain number of times that increases with n to ∞. As a result, the sequence $\{R^n\}$ cannot be normal at z_0, for if a subsequence

$\{R^\nu\}$ converges uniformly in Δ to a rational function $S(z)$ (which may be constant, finite or infinite), the equations $R^\nu(z) = \zeta$ for sufficiently large ν have the same fixed number of roots in Δ as does the equation $S(z) = \zeta$, as we saw in the proof of Theorem 5.2.6(ii). But this is a contradiction and shows that $J_R = \widehat{\mathbb{C}}$.

A more recent example is

$$R(z) = \left(\frac{z-2}{z}\right)^2,$$

given by R. Mañé, P. Sad, and D. Sullivan [1983]. The proof is based on the classification of components of the Fatou set and the fact that the critical points 0 and 2 are preperiodic (cf. also Beardon [1991], Theorem 9.4.4).

Entire Functions. The study of dynamical systems generated by entire functions goes back to Fatou [1926] and has had an analogous development to that for rational functions, although it was not until 1968 that Baker showed that the repulsive fixed points are dense in the corresponding Julia set. A major difference between the two theories is the *No Wandering Domains Theorem* of Sullivan for rational maps (cf. Beardon [1991], Chapter 8). The proof of the Baker result, given presently, makes compelling use of the Ahlfors theory of covering surfaces.

It is evident from Ahlfors Second Fundamental Theorem (§1.9) that if D_1, D_2, D_3 are suitable bounded domains in $\mathbb{C}$ with disjoint closures, and $f(z)$ is analytic in U, then, unless $f(z)$ maps an island in U univalently onto some D_j, we must have

$$S(r) < h_1 L(r),$$

where h_1 depends only on the D_j's. Then by Lemma 3.6.4, it follows that $f^\#(0) < h_2$, where h_2 depends only on h_1. We conclude that for some constant C depending only on the domains D_j and not on $f(z)$, if

$$f^\#(0) > C,$$

then $f(z)$ maps an island in U univalently onto some D_j. If $|z| < R$ is used instead of U, then for the latter conclusion we require $f^\#(0) > \frac{C}{R}$.

Theorem 5.2.11 *Let f be entire. Then the repulsive fixed points of $\{f^n\}$ are dense in J_f.*

Proof. As in the rational case,

$$\{\text{repulsive fixed points}\} \subset J_f.$$

Let $z_0 \in J_f$ and take an arbitrary neighbourhood N_{z_0} of z_0. We show that N_{z_0} contains a repulsive fixed point of some order. Since J_f is a perfect set, we may take three disjoint disks $D(a_i; \varepsilon) \subset N_{z_0}$ with $a_i \in J_f$, and such that none of their closures contain the exceptional point (!) if there is one.

Since $\{f^n\}$ is not normal in any neighbourhood of each a_i, we deduce by Marty's criterion (§3.3) that in each $D\left(a_i; \frac{\varepsilon}{3}\right)$, there is a point $b_i = b_i(n)$ such that

$$(f^n)^\#(b_i) > \frac{3C}{\varepsilon}, \qquad n > n_0(i), \tag{5.2}$$

where C is as above. Fix n so that (5.2) holds simultaneously in all three disks and define $g_i(\zeta) = f^n(z)$, where $z = b_i + \zeta$, $i = 1, 2, 3$. Then $g_i(\zeta)$ is analytic in $|\zeta| < \frac{\varepsilon}{3}$, which corresponds to the disk $D\left(b_i; \frac{\varepsilon}{3}\right)$. Since $g_i^\#(0) = (f^n)^\#(b_i)$, we obtain $\left(\text{with } R = \frac{\varepsilon}{3}\right)$ that $D\left(b_i; \frac{\varepsilon}{3}\right)$ contains an island which is mapped univalently by f^n onto some $D(a_j; \varepsilon)$. If we repeat this argument at most three times, we may conclude that some iterate of f^n, say f^ν, will have a simple island in $D\left(b_j; \frac{\varepsilon}{3}\right)$ over the disk $D(a_j; \varepsilon)$. Setting $\phi = (f^\nu)^{-1}$, and applying Rouché's theorem to the function

$$\phi(z) - z = \left(\phi(z) - a_j\right) - (z - a_j)$$

in $D(a_j; \varepsilon)$, leads to the conclusion that ϕ has a fixed point p in $D(a_j; \varepsilon)$, which we now call D.

Consider a Möbius transformation $L : \{|t| < 1\} \to D$ such that $L(0) = p$. Then $\Psi(t) = (L^{-1} \circ \phi \circ L)(t)$ maps $|t| < 1$ into $|t| < \sigma < 1$, $\Psi(0) = 0$. Thus the Schwarz Lemma applied to $\Psi(t)/\sigma$ yields $|\Psi'(0)| < \sigma < 1$. Since $\Psi'(0) = \phi'(p)$, we see that p is an attractive fixed point for $\phi = (f^\nu)^{-1}$, i.e., p is a repulsive fixed point for f^ν, and the theorem is proved.

Observe that the proof is readily adapted to the setting of rational functions.

Lastly we remark that Fatou in 1926 conjectured that $J_f = \mathbb{C}$ for $f(z) = e^z$. Baker [1970] showed that if $f(z) = kze^z$, for $k > 3e$, then $J_f = \mathbb{C}$, and indeed the conjecture has been resolved in the affirmative by Misiurewicz [1981]. Devaney [1984] has further shown that for $E_\lambda(z) = \lambda e^z$ and $\lambda > \frac{1}{e}$, then $J_{E_\lambda} = \mathbb{C}$.

5.3   Normal Functions

The concept of normal functions was introduced by Noshiro [1938] following from earlier related work of Yosida [1934]. Noshiro's definition was restricted to the unit disk and was later extended by Lehto and Virtanen in their seminal paper of 1957(a). In this work, the authors are interested in extending the Lindelöf property of analytic functions to meromorphic functions.

We say that a function $f(z)$ has the *Lindelöf perperty* in U if whenever $f(z) \to \alpha$ as $z \to z_0 \in \partial U$ along some arc lying in U and terminating at z_0, then $f(z) \to \alpha$ uniformly as $z \to z_0$ inside any angular domain of opening

$\pi - \varepsilon$ in U with z_0 as its vertex which is bisected by the radius drawn to z_0. In this case, $f(z)$ has the *angular limit* α at z_0

Although bounded analytic functions have the Lindelöf property (cf. Collingwood and Lohwater [1966], Theorem 2.3) a meromorphic function may not, and this led Lehto and Virtanen to the class of normal functions.

Definition 5.3.1 *A meromorphic function $f(z)$ is* **normal** *in a simply connected domain Ω if the family $\{f \circ t(z)\}$ is normal in Ω, where $t(z) = z'$ is an arbitrary one-to-one conformal mapping of Ω onto itself.*

Some examples of normal functions on a given simply connected domain are:

(i) If $f(z)$ is meromorphic and omits three values, then f is normal. For then the family $\{f \circ t(z)\}$ would omit the same three values and subsequently be normal. Thus the elliptic modular function $\mu(z)$ is normal.

(ii) If $f(z)$ is analytic and omits two values, then $f(z)$ is normal. Therefore, bounded analytic functions are normal.

(iii) If $f(z)$ is normal, then any branch of $\left(f(z)\right)^{\mu}$, $\mu \in \mathbf{R}$, is normal. If μ is not an integer, we assume $f(z) \neq 0, \infty$ in order that $\left(f(z)\right)^{\mu}$ be single-valued.

(iv) If $f(z)$ is normal meromorphic, then every rational function $R\left(f(z)\right)$ of $f(z)$ normal.

(v) If $f(z)$ is meromorphic, omits 0 and ∞, and takes a third value α only $(n-1)$ times, then $\left(f(z)\right)^{1/n}$ and $f(z)$ are normal. Indeed, the function $\left(f(z)\right)^{1/n}$ is single-valued, omitting 0, ∞ and one of the nth roots of α. By (ii), $\left(f(z)\right)^{1/n}$ is normal, and so is $\left(\left(f(z)\right)^{1/n}\right)^{n}$ by (iii).

We restrict our attention for the moment to the unit disk U. An important result of Noshiro [1938] characterizes normal functions in terms of the growth of their spherical derivatives.

Theorem 5.3.2 *A meromorphic function $f(z)$ in U is normal if and only if*

$$\sup_{z \in U}(1 - |z|^2)f^{\#}(z) < \infty, \qquad |z| < 1, \tag{5.3}$$

where $f^{\#}(z)$ is the spherical derivative.

Proof. If $f(z)$ is a normal function, then

$$\mathcal{F} = \left\{ f \circ t : t(z) = e^{i\gamma} \frac{z+a}{1+\bar{a}z}, \ \gamma \text{ real}, \ |a| < 1 \right\}$$

is an invariant normal family (cf. §3.6). By Theorem 3.6.2

$$g^{\#}(0) \leq M, \qquad g \in \mathcal{F},$$

and for $g = f \circ t$, this gives

$$f^{\#}(a) \leq \frac{M}{1 - |a|^2}.$$

On the other hand, if $f(z)$ is meromorphic satisfying (5.3), then the translates $f \circ t$ satisfy (3.3) of Theorem 3.6.2 and thus form an invariant normal family, that is, f is normal.

It is evident that every function belonging to an invariant normal family is normal, and conversely every normal function belongs to an invariant normal family. In particular, if f is a normal function, $T_0(r, f) < C \log\left(1/(1-r^2)\right)$ (Theorem 3.6.7).

Theorem 5.3.2 provides yet another rather important class of normal functions.

(vi) Any analytic univalent function is normal.

Proof. As we may take the domain to be the unit disk U, note that any analytic univalent function $f(z)$ is expressible in the form $f(z) = \alpha g(z) + \beta$, $\alpha, \beta \in \mathbb{C}$, where g belongs to the schlicht class $\mathcal{S}$ of §1.4, no. 5. Then one can write

$$\frac{(1-|z|^2)|f'(z)|}{1+|f(z)|^2} = \frac{(1-|z|^2)|\alpha g(z)|}{1+|\alpha g(z)+\beta|^2} \cdot \left| \frac{g'(z)}{g(z)} \right|,$$

which remains bounded as $|z| \to 1$ owing to the standard estimate for schlicht functions (cf. Pommerenke [1975], Theorem 1.6)

$$\left| z \frac{g'(z)}{g(z)} \right| \leq \frac{1+|z|}{1-|z|}.$$

This proves (vi).

In Example 2.2.7, we deduced that for f analytic in U,

$$(1-|z|)|f'(z)| \leq \sqrt{\frac{1}{\pi} D_U(f)}, \qquad |z| < 1,$$

where $D_U(f)$ is the Dirichlet integral of f. Therefore, if $f \in AD(U)$, the space of analytic Dirichlet-finite functions on U, then

$$\sup_{z \in U} (1-|z|^2) \frac{|f'(z)|}{1+|f(z)|^2} \leq 2\sqrt{\frac{1}{\pi} D_U(f)} < \infty,$$

that is,

(vii) $f \in AD(U)$ implies f is a normal function in U, and what is more, f is a Bloch function.

If $f(z)$ is merely analytic in U and satisfies the condition

$$\sup_{z \in U}(1 - |z|^2)|f'(z)| < \infty,$$

then $f(z)$ is termed a *Bloch function*. From Theorem 5.3.2 it is evident that *every Bloch function is normal*. Moreover, if $r(f) = \sup_{a \in U} r(a, f)$ is defined as in §4.3, then $f(z)$ is a Bloch function if and only if $r(f) < \infty$ (cf. Pommerenke [1970]).

A class of meromorphic functions called *absolutely hypernormal* are shown to coincide with the Bloch functions in Anderson and Rubel [1978].

Theorem 5.3.2 can, of course, be rephrased in terms of the Poincaré metric $d\sigma$; that is, the normality of $f(z)$ is U is equivalent to

$$f^{\#}(z)|dz| \leq M d\sigma, \qquad M > 0.$$

Moreover, since $d\sigma$ and $f^{\#}(z)|dz|$ are invariant under one-to-one conformal mappings, we may conclude (Lehto and Virtanen [1957a]):

Corollary 5.3.3 *A meromorphic function $f(z)$ is normal in a simply connected domain D of hyperbolic type (i.e., conformally equivalent to U) if and only if*

$$f^{\#}(z)|dz| \leq M d\sigma, \qquad z \in D,$$

for some constant M.

Condition (5.3) has also been reformulated in another way to provide the following criterion for a function to be normal. This is the theorem of Lohwater and Pommerenke [1973] upon which the Zalcman Lemma (§4.1) is based, and since their proofs are nearly the same we omit the details.

Theorem 5.3.4 *A nonconstant meromorphic function f in U is normal if and only if there do not exist sequences $\{z_n\} \subseteq U$ and $\{\rho_n\}$ with $\rho_n > 0$, $\rho_n \to 0$, such that*

$$f(z_n + \rho_n \zeta) \to g(\zeta),$$

as $n \to \infty$, spherically uniformly on compact subsets of $\mathbb{C}$, where g is a nonconstant meromorphic function in $\mathbb{C}$.

Finally there is the Lehto-Virtanen extension of the Lindelöf property.

Theorem 5.3.5 *If $f(z)$ is meromorphic and normal in U and has asymptotic value α at a boundary point z_0 taken along a Jordan curve lying in U, then $f(z)$ possess the angular limit α at z_0.*

The proof of this notable result is based on the following phenomenon (Lehto and Virtanen [1957a], Theorem 1; cf. also Noshiro [1960], p. 86).

Theorem 5.3.6 *Let $f(z)$ be meromorphic in U, and suppose that $f(z)$ has the asymptotic value zero along a Jordan curve lying in U and terminating at $z_0 \in \partial U$, but that $f(z)$ does not have the angular limit zero at z_0. Then, if $m > 0$, there exists a Jordan curve L in U terminating at z_0 on which $f(z)$ tends to zero as $z \to z_0$, and a sequence $\{z_n\} \subseteq U$, $z_n \to z_0$, at which $f(z_n) = a \neq 0$ for all n, and such that all the points z_n have hyperbolic distance less than m from L.*

To see how Theorem 5.3.5 may be deduced from this, let $f(z)$ be meromorphic and normal in U, and without loss of generality assume the asymptotic value $\alpha = 0$. Contrary to the assertion, suppose that $f(z)$ does not have angular limit zero at z_0. Invoking the preceding theorem yields the Jordan curve L and sequence of points $\{z_n\}$ as given therein. Choose a sequence of one-to-one conformal maps $S_n : U \to U$ satisfying $S_n(0) = z_n$. Denote by Δ the closed hyperbolic disk with hyperbolic centre 0, of hyperbolic radius $m+1$. Then Δ is mapped by S_n onto a similar such disk having hyperbolic centre z_n, with L passing interior to each. Since $\{f \circ S_n\}$ is normal in U, we can extract a convergent subsequence $\{f \circ S_{n_k}\}$ converging normally to a function ϕ, which is either meromorphic or $\equiv \infty$.

On the other hand, as the hyperbolic metric is invariant under one-to-one conformal mappings, each inverse function $S_{n_k}^{-1}$ maps an arc or arcs of L into the interior of Δ, Δ^0. Then for large values of k, the functions $f\big(S_{n_k}(z)\big)$ take small values on these image arcs because $f(z) \to 0$ along L (in the direction of z_0). As a consequence, ϕ is meromorphic. Moreover, the image arcs of L in Δ^0 under the mappings $S_{n_k}^{-1}$ must have an accumulation continuum $\gamma \subseteq \Delta$, and we deduce that $\phi(z) = 0$ on γ. Therefore, $\phi \equiv 0$ in U. This contradicts $\phi(0) = a \neq 0$, and completes the proof of Theorem 5.3.5.

Further developments on normal functions can be found in Lappan [1961], Bagemihl [1961, 1963], Bagemihl and Seidel [1960, 1961], Gavrilov[1961a, 1961b], Anderson, Clunie, Pommerenke [1974], amongst others.

5.4 Harmonic Functions

A *harmonic* function $u(x,y)$ is a C^2-solution of the Laplace equation

$$\Delta u = \frac{\partial^2 u}{\partial x^2} + \frac{\partial^2 u}{\partial y^2} = 0.$$

Throughout this section we shall use the notation $u(x,y)$, $u(z)$, $u(re^{i\theta})$, and $u(r,\theta)$ interchangeably, where $z = x + iy = re^{i\theta}$.

If $f(z) = u(x,y) + iv(x,y)$ is analytic, then $u = u(x,y)$ is harmonic and $v = v(x,y)$ is the *harmonic conjugate* of u, with u and v satisfying the

Cauchy-Riemann equations

$$\frac{\partial u}{\partial x} = \frac{\partial v}{\partial y}, \quad \frac{\partial u}{\partial y} = -\frac{\partial v}{\partial x}.$$

Whenever u is harmonic in a simply connected domain Ω, an analytic function f can always be determined so that $u = \mathcal{R}e\, f$ in Ω. To see this, fix z_0 in Ω, and define

$$f(z) = u(z_0) + \int_\gamma F(\zeta)\, d\zeta,$$

where $F = u_x - iu_y$, and γ is a contour in Ω from z_0 to z. Since F is C^1 and its real and imaginary parts satisfy the Cauchy-Riemann equations, F is analytic in Ω, so that by direct calculation, $f(z)$ is analytic in Ω with $f'(z) = F(z)$. If we set $f = \mathcal{U} + i\mathcal{V}$, then

$$(\mathcal{U} - u)_x = (\mathcal{U} - u)_y = 0,$$

implying $\mathcal{U} - u = $ constant in Ω. Since $f(z_0) = u(z_0) = \mathcal{U}(z_0)$, we have $\mathcal{R}e\, f = \mathcal{U} = u$, as required.

Given a continuous function f on the circle $|z| = R$, the *Poisson formula*

$$u(z) = \frac{1}{2\pi} \int_0^{2\pi} \frac{R^2 - |z|^2}{|\zeta - z|^2} f(Re^{i\phi})\, d\phi, \qquad \zeta = Re^{i\phi}, \tag{5.4}$$

represents a harmonic function in $|z| < R$ with $\lim_{z \to \zeta} u(z) = f(\zeta)$. Moreover, if u is harmonic in $|z| < R$ and continuous on $|z| \le R$, then u has the Poisson representation (5.4) in terms of its boundary values. When the point z is at the centre of the disk, (5.4) reduces to the *mean value property*

$$u(0) = \frac{1}{2\pi} \int_0^{2\pi} u(Re^{i\phi})\, d\phi.$$

This can also be expressed in terms of an *areal mean*

$$u(0) = \frac{1}{\pi R^2} \int_0^{2\pi} \int_0^R u(r, \phi) r\, dr\, d\phi.$$

A continuous function u on a domain Ω which satisfies the mean value property for all sufficiently small disks contained in Ω is necessarily harmonic (cf., e.g., Helms [1969], Corollary 2.9).

We may conclude from the preceding that if $\{u_n\}$ is a sequence of harmonic functions converging uniformly on compact subsets of a domain Ω to a function u, then u is harmonic in Ω.

There is also a *Poisson-Stieltjes representation* for a positive harmonic function u in $|z| < R$ (Herglotz [1911]),

$$u(z) = \frac{1}{2\pi} \int_0^{2\pi} \frac{R^2 - |z|^2}{|\zeta - z|^2}\, d\mu(\phi), \qquad \zeta = Re^{i\phi}, \ |z| < R,$$

where $\mu(\phi)$ is a nondecreasing function on $[0, 2\pi]$. Conversely, any such representation denotes a positive harmonic function in $|z| < R$.

Another consequence of the formula (5.4) is the well-known

Harnack's Inequality *If u is a positive harmonic function on Ω and K is a compact subset of Ω, then there exists a constant $\kappa = \kappa(\Omega, K)$ such that*

$$u(z) \leq \kappa\, u(\zeta),$$

for all $z, \zeta \in K$.

Observe that the constant κ is independent of the harmonic function u.

Definition 5.4.1 *A family $\mathcal{H}$ of harmonic functions is **normal** in a domain Ω if every sequence $\{u_n\} \subseteq \mathcal{H}$ contains a subsequence which converges uniformly on compact subsets of Ω to either a harmonic function or to $\pm\infty$.*

Unfortunately, the real parts of a normal family of analytic functions do not necessarily constitute a normal family of harmonic functions. For instance, if $z \in U$, define

$$\begin{aligned} f_n(z) &= nx + i(ny + n^2), \qquad n = 1, 2, 3, \ldots, \\ &= u_n(x, y) + i v_n(x, y), \end{aligned}$$

i.e.,

$$f_n(z) = nz + in^2.$$

Then $f_n(z)$ is analytic and $f_n \to \infty$ uniformly in U, so that $\{f_n\}$ is normal in U. However, this is certainly not the case for the family of harmonic functions $\{u_n(x, y) = nx : n = 1, 2, 3, \ldots\}$.

Nevertheless, it is evident that: if $\mathcal{F}$ is a normal family of analytic functions which do not admit ∞ as a limit, then $\mathcal{H} = \mathcal{R}e\,\mathcal{F}$ is a normal family of harmonic functions.

On the other hand, if $\mathcal{H} = \{u\}$ is a normal family of harmonic functions in a simply connected domain Ω, and $\mathcal{F}$ is the corresponding family of analytic functions $\{f\}$ in Ω such that $\mathcal{R}e\,f = u$, then $\mathcal{F}$ is normal. In fact, take $\{f_n\} \subseteq \mathcal{F}$, where $\mathcal{R}e\,f_n = u_n \in \mathcal{H}$. First, suppose that $\{u_n\}$ has a convergent subsequence, again denoted by $\{u_n\}$, converging normally to $\pm\infty$. Since $|f_n| \geq |u_n|$, the sequence $\{f_n\}$ converges normally to ∞. Next, suppose that the subsequence $\{u_n\}$ converges normally in Ω to a harmonic function u. Then $\{u_n\}$ is locally bounded, and hence so is $\{f_n'\}$, since

$$f_n' = \frac{\partial u_n}{\partial x} - i \frac{\partial u_n}{\partial y}$$

(cf. Proposition 5.4.8). Consequently, $\{f_n\}$ has a normally convergent subsequence by Theorem 2.2.6 (or Marty's theorem), proving the result.

In common with families of analytic functions there is the harmonic analogue of Montel's theorem (cf. Koebe [1909], or Helms [1969], Theorem 2.18, for a direct proof via the Arzelà–Ascoli Theorem):

Theorem 5.4.2 *A locally bounded family $\mathcal{H}$ of harmonic functions on a domain Ω is normal.*

Proof. If suffices to show that the family $\mathcal{H}$ is normal at each point of Ω since normality in a domain is equivalent to pointwise normality just as in the analytic case (cf. Theorem 2.1.2). Let $\{h_n\}$ be a sequence in $\mathcal{H}$ and z_0 an arbitrary point of Ω, with $D = D(z_0; r) \subseteq \Omega$ a disk in which $|h_n(z)| \leq M$ for $n = 1, 2, 3, \ldots$. Then there are analytic functions f_n in D such that $h_n = \mathcal{R}e\, f_n$, so that the analytic function $F_n = e^{f_n}$ satisfies

$$e^{-M} \leq |F_n| = e^{h_n} \leq e^M.$$

By virtue of Montel's theorem, there is a subsequence $\{F_{n_k}\}$ satisfying $F_{n_k} \to F$ normally in D and F is analytic there. Since $e^{-M} \leq |F| \leq e^M$, the uniform continuity of the log function on $[e^{-M}, e^M]$ implies

$$h_{n_k} = \log|F_{n_k}| \to \log|F| = h,$$

normally in D, where h is harmonic. This completes the proof.

The family $\mathcal{H}$ is clearly compact as well.

From the normality criterion of boundedness, the weaker criteria of being bounded above or bounded below can be derived. Firstly,

Theorem 5.4.3 *The family $\mathcal{H}^+$ of positive harmonic functions on a domain Ω is normal.*

Proof. Again it suffices to show that $\mathcal{H}^+$ is normal in a neighbourhood of an arbitrary point $z_0 \in \Omega$. Given a sequence $\{u_n\} \subseteq \mathcal{H}^+$, choose a subsequence, again denoted by $\{u_n\}$, for which $u_n(z_0)$ converges to a limit $L \leq \infty$. Then in some closed disk $K(z_0; r) \subseteq \Omega$, an application of the Harnack Inequality with $u_n = u$ gives

$$\frac{1}{\kappa} u_n(z_0) \leq u_n(z) \leq \kappa\, u_n(z_0), \qquad z \in K(z_0; r).$$

If $L < \infty$, then $\{u_n(z)\}$ is bounded, hence normal in a neighbourhood of z_0. If $L = \infty$, then $u_n \to \infty$ uniformly on $K(z_0; r)$, and normality at z_0 is again achieved.

Corollary 5.4.4 *If $\mathcal{H}^+$ is a family of positive harmonic functions in Ω such that $u(z_0) = a_0 \in \mathbf{R}$ for all $u \in \mathcal{H}^+$, then $\mathcal{H}^+$ is normal and compact.*

Furthermore, it is evident from the theorem that a family of harmonic functions which is bounded above by M or bounded below by m, is normal by considering the functions $M - u$ or $u - m$, respectively. Another immediate consequence is

Harnack's Principle *Let $\{u_n\}$ be a sequence of harmonic functions in a domain Ω which satisfy $u_{n+1} \geq u_n$, $n = 1, 2, 3 \ldots$. Then either $u_n \to +\infty$ uniformly on compact subsets of Ω or $\{u_n\}$ converges to a harmonic function, uniformly on compact subsets.*

For, the functions $u_n - u_1$, $n = 1, 2, 3, \ldots$, are nonnegative harmonic and as such form a normal family. The sequence $\{u_n - u_1\}$ is also monotone and either converges or diverges to $+\infty$ at each point of Ω. By the normality, this sequence either converges normally to a harmonic function or to $+\infty$ in Ω, and likewise for the sequence $\{u_n\}$.

An analogue of the FNT (§2.7) is now readily deduced.

Corollary 5.4.5 *A family $\mathcal{H}$ of harmonic functions in a domain Ω which omit one specific real value α is normal.*

In fact, for each $u \in \mathcal{H}$, either $u > \alpha$ or $u < \alpha$ in Ω. This is so because if $u(z_1) > \alpha$ and $u(z_2) < \alpha$, $z_1, z_2 \in \Omega$, then u would assume the value α at some point z on each curve in Ω joining z_1 and z_2. Thus $\mathcal{H}$ can be partitioned into two subfamilies

$$\mathcal{H}_1 = \{u \in \mathcal{H} : u > \alpha\}, \quad \mathcal{H}_2 = \{u \in \mathcal{H} : u < \alpha\}.$$

The normality of $\mathcal{H}_1$ and $\mathcal{H}_2$ is then deduced, and so $\mathcal{H}$ is normal.

This result has been extended to solutions of an elliptic partial differential equation by Beardon [1971].

In the same manner one can demonstrate that: *The family $\mathcal{H}$ of harmonic functions which do not take all of the values in the interval $(0, 1)$ forms a normal family.* For, each such function u must satisfy either $u > 0$ or $u < 1$ in Ω, and then proceed as above.

As in the analytic case (Corollary 2.2.4), *a normal family of harmonic functions which is bounded at a point is locally bounded.* This observation permits harmonic analogues of the Schottky and Landau theorems of §2.8 (cf. Montel [1935]).

Theorem 5.4.6 *Let $\mathcal{H} = \{u\}$ be a normal family of harmonic functions in $|z| < R$ which satisfy $u(0) = a_0$. Then for each fixed θ, with $0 < \theta < 1$, there is a constant $M = M(a_0, \theta)$ such that*

$$|u(z)| \leq M(a_0, \theta), \qquad |z| \leq \theta R, \ u \in \mathcal{H}.$$

The result is proved by considering the functions $U(z) = u(Rz)$, $|z| < 1$, as in the analytic case.

A harmonic analogue of Landau's theorem is more challenging. Consider the functions $\{u\}$ harmonic in $|z| < R$ satisfying $u(0) = a_0$. Further suppose that each u satisfies

$$\frac{\partial u}{\partial x} = a_1 \neq 0$$

at the origin, and so has the representation

$$u(x, y) = a_0 + a_1 x + b_1 y + \ldots .$$

Finally, assume that $\{u\}$ possesses some criterion of normality that depends solely on the values of each function u, such as being positive or omitting a given value.

The corresponding functions

$$U(x, y) = u(Rx, Ry) = a_0 + a_1 Rx + b_1 Ry + \ldots$$

are harmonic in $|z| < 1$ and also form a normal family there. Then for $|z| \leq \frac{1}{2}$,

$$|U(z)| \leq M = M\left(a_0, \tfrac{1}{2}\right),$$

by the preceding theorem. Let $K = K(z_0; r)$ be a closed disk in $|z| \leq \frac{1}{2}$. Applying the areal mean value property to $\frac{\partial U}{\partial x}$ and then Green's formula gives

$$\left(\frac{\partial U}{\partial x}\right)_{z_0} = \frac{1}{\pi r^2} \int \int_K \frac{\partial U}{\partial x} dx\, dy$$

$$= \frac{1}{\pi r^2} \int_{|z-z_0|=r} U\, dy = \frac{1}{\pi r^2} \int_0^{2\pi} U r \cos\theta\, d\theta.$$

Therefore,

$$\left|\left(\frac{\partial U}{\partial x}\right)\right|_{z_0} \leq \frac{2M}{r}.$$

In particular, the partial derivative $a_1 R$ at the origin, with $r = \frac{1}{2}$, yields the inequality

$$|a_1| R \leq \frac{2M\left(a_0, \frac{1}{2}\right)}{\frac{1}{2}},$$

i.e.,

$$R \leq \frac{4M\left(a_0, \frac{1}{2}\right)}{|a_1|}.$$

The preceding analysis can then be summarized as

Theorem 5.4.7 *Let $\mathcal{H}$ be a family of harmonic functions possessing a criterion of normality in $|z| < R$ such as stipulated above, and such that each u in $\mathcal{H}$ satisfies $u(0) = a_0$, $\left(\frac{\partial u}{\partial x}\right)_{z=0} = a_1 \neq 0$. Then there exists a number $R(a_0, a_1) > 0$ for which the following holds: any function $h(x, y) = a_0 + a_1 x + b_1 y + \ldots$ which is harmonic in a disk about the origin of radius greater than $R(a_0, a_1)$ cannot satisfy the normality criterion of $\mathcal{H}$.*

Clearly, the condition $\left(\frac{\partial u}{\partial y}\right)_{z=0} = b_1 \neq 0$ may also be used.

We have essentially demonstrated something further (compare Corollary 2.2.5).

Proposition 5.4.8 *Let $\mathcal{H}$ be a normal family of harmonic functions in Ω such that $|u(\zeta_0)| \leq m$ for each $u \in \mathcal{H}$, where ζ_0 is a fixed point of Ω. Then $\left\{\frac{\partial u}{\partial x}\right\}$ is locally bounded, $u \in \mathcal{H}$.*

Indeed, $\mathcal{H}$ is locally bounded, so that

$$|u(z)| \leq M_1, \qquad u \in \mathcal{H},$$

whenever $z \in K = K(z_0; r_0)$, say. Applying the areal mean value property to $\frac{\partial u}{\partial x}$ together with Green's formula as above yields the result.

Bloch Principle. From what has already transpired, it should be apparent that there is a Bloch Principle operating for harmonic functions. Specifically, *any property $\mathcal{P}$ which reduces a function which is harmonic in $\mathbb{C}$ to a constant is likely to make a family of harmonic functions possessing $\mathcal{P}$ normal.* A rigorous version, along the lines of §4.1, is left as an exercise for the reader. Instances already encountered include the properties of boundedness, positivity, and omitting a given real value. One further illustration is

Example 5.4.9 We define

$$\mathcal{H}_M = \{u \text{ harmonic in } U : D_U(u) \leq M < \infty\},$$

where

$$D_U(u) = \int\int_U |\operatorname{grad} u|^2 \, dx \, dy$$

is the Dirichlet integral of u (cf. Example 2.2.7 for the analytic counterpart). Since $u = \mathcal{R}e \, f$ for $f(z)$ analytic in U, we have

$$D_U(u) = D_U(f) = \int\int_U |f'(z)|^2 \, dx \, dy.$$

Therefore, given $u \in \mathcal{H}_M$, the analysis of Example 2.2.7 then implies that $\{f' : u = \mathcal{R}e \, f, u \in \mathcal{H}_M\}$ is locally bounded in U. By taking the real parts appropriately in cases (i) and (ii) of Theorem 2.2.6, we find that $\mathcal{H}_M$ is normal in U.

This result is related to Lebesgue's 1907 work on the Dirichlet Principle, a proof of which, based upon Example 5.4.9, can be found in Tsuji [1959], pp. 12–14.

The foregoing example may be rephrased: *If $\{f'\}$ is locally bounded, then $\{u = \mathcal{R}e\, f\}$ is normal.* Thus, if the harmonic functions of $\mathcal{H} = \{u\}$ have locally bounded partial derivatives, u_x, u_y, then $\mathcal{H}$ is normal.

On the other hand, if u is harmonic in $\mathbb{C}$ and $D_{\mathbb{C}}(u) < \infty$, then u must be constant. To see this, define $f = u + iv$ so that f is analytic in $\mathbb{C}$. Then, given $r > 0$, the areal mean value property followed by Schwarz's inequality yields

$$|\pi r^2 f'(0)|^2 = \left| \int\int_{|z|<r} f'(z)\, dx\, dy \right|^2$$
$$\leq \left(\int\int_{|z|<r} |f'(z)|^2\, dx\, dy \right) \left(\int\int_{|z|<r} dx\, dy \right)$$
$$\leq \pi r^2 D_{\mathbb{C}}(u).$$

Therefore $f'(0) = 0$, and similarly $f'(z_0) = 0$ for any $z_0 \in \mathbb{C}$. Hence, f is identically constant, and so u must be constant. Thus we have gleaned another example of the Bloch Principle.

However, there are exceptions to the Bloch Principle as in the analytic case. Indeed, let us consider a variant of Counterexample 4.2.2.

Let $\mathcal{P}$ be the property of the harmonic function u that *the differential expression*

$$(u_x - 1)(u_y - 1)(u_x + u_y - u)$$

omits the value zero. If u is harmonic in $\mathbb{C}$ and possesses $\mathcal{P}$, then u_x, u_y are harmonic functions in $\mathbb{C}$ omitting the value 1, so that u_x, u_y are constant (by Picard's little theorem). Hence $u(x,y) = ax + by + c$. Moreover, as $(u_x + u_y - u) \neq 0$, it must be the case that $a = b = 0$, and consequently u is constant.

For the other part of the argument, consider the family

$$\mathcal{H} = \{u_n(x,y) = nx + ny : n = 2,3,4,\ldots\}$$

for $z = x + iy$ in the unit disk U. Then it is verified that

$$\left(\frac{\partial u_n}{\partial x} - 1\right)\left(\frac{\partial u_n}{\partial y} - 1\right)\left(\frac{\partial u_n}{\partial x} + \frac{\partial u_n}{\partial y} - u_n\right) > 0, \qquad n = 2,3,4,\ldots,$$

in U, that is, each $u_n \in \mathcal{H}$ satisfies the property $\mathcal{P}$. However, $\mathcal{H}$ is not a normal family in U, and the Bloch Principle is violated.

Furthermore, let us now consider the converse; cf. Theorem 4.1.6 for the analytic counterpart. Following §4.1, we consider a property $\mathcal{P}$ of harmonic functions which satisfies the conditions

(i) If $\langle u, \Omega \rangle \in \mathcal{P}$ and $\Omega' \subseteq \Omega$, then $\langle u, \Omega' \rangle \in \mathcal{P}$.

(ii) If $\langle u, \Omega \rangle \in \mathcal{P}$ and $\phi(z) = az + b$, then $\langle u \circ \phi, \phi^{-1}(\Omega) \rangle \in \mathcal{P}$.

Theorem 5.4.10 *Let $\mathcal{P}$ be a property satisfying (i) and (ii) above. Assume that if $\mathcal{H}$ is any family of harmonic functions defined in U with $\langle u, U \rangle \in \mathcal{P}$, for all $u \in \mathcal{H}$, then $\mathcal{H}$ is necessarily normal. Then any u_0 which is harmonic in $\mathbb{C}$ with $\langle u_0, \mathbb{C} \rangle \in \mathcal{P}$, reduces to a constant.*

The proof follows a similar line of reasoning as in Theorem 4.1.6 with Proposition 5.4.8 in place of Marty's theorem. Details are left to the reader.

As a consequence, a nonconstant harmonic function in $\mathbb{C}$ cannot be bounded, positive, admit one exceptional value, or have finite Dirichlet integral.

Julia's Theorem. Another result worth discussing that has a harmonic analogue is the classical Julia Theorem (§2.8).

Let u be a nonconstant harmonic function in $\mathbb{C}$ and set

$$U_n(x, y) = u(2^n x, 2^n y)$$

for $z = x + iy$ belonging to the unit disk. Observe that the family $\{U_n\}$ cannot be normal in U for, if it were, the condition

$$U_n(0) = u(0) = a_0$$

would compel the functions U_n to be bounded on, say, the disk $|z| \le \frac{1}{2}$. Consequently, u would be bounded in $\mathbb{C}$, and hence constant. It follows that there is an *irregular point* in the unit disk at which $\{U_n\}$ in not normal. It is evident that $z = 0$ could not be the only irregular point in U since if this were the case, a subsequence of $\{U_n\}$ which converged uniformly on $|z| = r < 1$ would do likewise in $|z| \le \rho < r$, by the Poisson formula. This also demonstrates, incidentally, that an irregular point in the present context cannot be isolated, and since the set of irregular points is closed, it is a *perfect set* (cf. also Lavrentieff [1927]).

There is, therefore, a point $\zeta_0 \ne 0$ such that in an arbitrarily small disk about ζ_0 the family $\{U_n\}$ is not normal. This means that $\{U_n\}$ attains every value in such a disk $\Delta_0 : |z - \zeta_0| < r$ for infinitely many n. Determine a sequence of homothetic disks

$$\Delta_n = \{|z - 2^n \zeta_0| < 2^n r\}, \qquad n = 1, 2, 3, \ldots,$$

so that $U_n(\Delta_0) = u(\Delta_n)$. Then u must take every value $\alpha \in \mathbb{R}$ in the family of disks $\{\Delta_n\}$ for infinitely many n; indeed, α is attained infinitely many times. We have established

Julia's Theorem *If u is a nonconstant harmonic function in the complex plane, then there is at least one ray* $\arg z = \theta$ *emanating from the origin, such that in every sector* $\theta - \varepsilon < \arg z < \theta + \varepsilon$, $u(z)$ *assumes, infinitely often, every real value.*

With regard to an entire function $f(z) = u(z) + iv(z)$, a line of Julia of $f(z)$ is also one for $u(z)$ $\big($and $v(z)\big)$, but not conversely. For instance, if

$$f(z) = e^z = e^x \cos y + ie^x \sin y,$$

the lines of Julia for $f(z)$ are the positive and negative imaginary axis. However, the lines of Julia for $u(z) = e^x \cos y$ are all lines issuing from the origin in the half-plane $\mathcal{R}e\, z \geq 0$.

Further investigations, in the spirit of Picard's second theorem, have been carried out by Montel [1935].

Hardy – Montel Theorem. To conclude our discussion of harmonic functions, we endeavour to prove an analogue of Montel's theorem (Theorem 2.9.3) due to Hardy [1926]. A direct analogue is not possible as the harmonic function $u(x, y) = x$ tends to different limits on different lines. Moreover, the function $u(x, y) = \sinh x \sin y$ tends to a limit on only one line. In spite of these differences, Hardy was able to demonstrate the following version of Montel's theorem, in a manner keeping in spirit with the original proof. The domain in this instance is taken to be the whole infinite strip $S = \{x + iy : \alpha < x < \beta\}$.

Hardy – Montel Theorem *Let* $u(x, y)$ *be a bounded harmonic function in the strip S and suppose that* $u(a, y) \rightarrow A$, *and* $u(b, y) \rightarrow B$ *as* $y \rightarrow +\infty$, *for* $\alpha < a < b < \beta$. *Then*

$$u(x, y) \rightarrow \frac{b - x}{b - a}A + \frac{x - a}{b - a}B,$$

uniformly in x in any strip interior to S.

Proof. We can reduce the problem to the special case of $A = B = 0$ by considering the harmonic function

$$v(x, y) = u(x, y) - \left\{ \frac{b - x}{b - a}A + \frac{x - a}{b - a}B \right\}.$$

Furthermore, we may suppose that $a = 0$, $b = \pi$. Then the "Poisson formula" representation for u in S (which can be derived, after first making the transformation $Z = e^{iz}$, from the upper half-plane representation; cf. Ahlfors [1979], p. 171) is given by

$$u(x, y) = \frac{1}{2\pi} \int_{-\infty}^{\infty} \frac{\sin x \, u(0, \eta) d\eta}{\cosh(\eta - y) - \cos x} + \frac{1}{2\pi} \int_{-\infty}^{\infty} \frac{\sin x \, u(\pi, \eta) d\eta}{\cosh(\eta - y) + \cos x},$$

$0 \leq x \leq \pi$, from which it follows that $u(x,y) \to 0$ as $y \to +\infty$, for $a \leq x \leq b$.

Next, suppose that $u = \mathcal{R}e\, f$ and let

$$g(z) = e^{-f(z)}, \qquad z \in S.$$

Then

$$e^{-2u} = |g|^2 = g(x+iy)\bar{g}(x-iy),$$

where $\bar{g}(x-iy) = \overline{g(x+iy)}$. Clearly g is bounded in S with $|g| \to 1$ whenever $a \leq x \leq b$ and $y \to +\infty$.

At this stage we replace the real variable x by the complex variable $\zeta = \xi + i\eta$ and define

$$G_y(\zeta) = G(\xi+i\eta) = g(\xi+i\eta+iy)\bar{g}(\xi+i\eta-iy).$$

The result is that $G_y(\zeta)$ is analytic and bounded in the rectangle

$$R: \alpha + \frac{1}{2}\delta \leq \xi \leq \beta - \frac{1}{2}\delta, \qquad -\delta \leq \eta \leq \delta \ \ (0 < 2\delta < \beta - \alpha).$$

If ζ is any point on the interval (a,b), interior to R, then $G_y(\zeta) \to 1$ as $y \to +\infty$. In view of the Vitali-Porter Theorem (§2.4), $G(\zeta) \to 1$ uniformly in any rectangle interior to R, and consequently $u \to 0$ uniformly in the strip $(\alpha+\delta, \beta-\delta)$, as $y \to +\infty$.

In a similar manner, one can derive the bounded analytic version of Montel's theorem by considering the uniformly bounded sequence of analytic functions

$$f_\eta(z) = f(x+iy+i\eta),$$

for $z = x + iy$ in the strip S. In fact, if $f_\eta(x_0+iy) \to \ell$ as $\eta \to \infty$ at the points $x_0 + iy$ lying in a rectangle R interior to S, then an application of Vitali-Porter again gives the desired uniform convergence of f to ℓ in any strip interior to S.

In contrast with Hardy's result, a Montel-type theorem is attainable in the half-strip $S : \alpha < x < \beta, y > 0$, in the following sense (suggested by W. Hayman): *If $u(x,y)$ is harmonic in S, $u(x,y) < M$, $u(x_0,y) \to M$ as $y \to \infty$, then $u(x,y) \to M$ as $y \to \infty$ uniformly for $\alpha+\delta \leq x \leq \beta-\delta$.* This can be seen by finding an analytic function f in S with $u = \mathcal{R}e\, f$, setting $g = e^f$ and applying the theorem of Cartwright quoted at the end of §2.9 to the function g.

Some of the normal family theory of harmonic functions can be developed for *subharmonic functions* and, in turn, for solutions to certain elliptic partial differential equations. In particular, this has been done in the case of Montel's theorem (Theorem 2.9.3) by Beardon [1971].

Martin Kernel. R.S. Martin, in his celebrated paper of 1941, demonstrated an integral representation of positive harmonic functions "bearing certain features of analogy with the Poisson-Stieltjes formula". Starting with a domain R, a compactification R_M^* is generated, where the set $\Delta_M = R_M^* - R$ is the *Martin boundary*. It is then shown that every positive harmonic function on R has the integral representation

$$h(z) = \int_{\Delta_M} K(z, \zeta) \, d\mu(\zeta), \qquad z \in R, \tag{5.5}$$

where μ is a measure on Δ_M depending on h and $K(z, \zeta)$ is the *Martin kernel*.

The existence of the kernel function arises from an application of Corollary 5.4.4. To see this, let $G(z, \zeta)$ be the generalized Green's function for R, that is, for fixed z, $G(z, \zeta)$ is positive harmonic in ζ except at $\zeta = z$ where it has a logarithmic singularity, is symmetric in (z, ζ), and approaches zero at all regular boundary points of ∂R and at ∞ if R is unbounded. Define for z, $\zeta \in R$, $\zeta_0 \in R$ (fixed)

$$K(z, \zeta) = \begin{cases} \dfrac{G(z, \zeta)}{G(z, \zeta_0)} & \text{if } z \neq \zeta_0 \\ 0 & \text{if } z = \zeta_0, \ \zeta \neq \zeta_0 \\ 1 & \text{if } z = \zeta = \zeta_0. \end{cases}$$

Then for fixed z, the function $K(z, \zeta)$ is a positive harmonic function of ζ except at $\zeta = z$, and $K(z, \zeta_0) = 1$.

Next, let $\{z_n\}$ be a sequence of points in R having no accumulation point in R. If R' is any open subset of R with compact closure $\overline{R'} \subseteq R$, and $\zeta_0 \in R'$, then the functions $K(z_n, \zeta)$ are positive harmonic functions of ζ in R' for all sufficiently large n. By Corollary 5.4.4 we deduce that there is a subsequence converging normally in R' to a positive harmonic function. The corresponding subsequence of $\{z_n\}$ is called a *fundamental sequence*. Two fundamental sequences are *equivalent* (and form an equivalence relation) if their corresponding $K(z, \zeta)$'s have the same limit function.

Definition 5.4.11 *Each equivalence class of fundamental sequences of R determines an* **ideal boundary element** *of R, the set of which constitute the Martin boundary Δ_M.*

Thus, for $z \in \Delta_M$, we can write the Martin kernel as

$$K(z, \zeta) = \lim_{n \to \infty} K(z_n, \zeta), \qquad \zeta \in R,$$

where $\{z_n\}$ is any fundamental sequence which determines z. If $z \in \Delta_M$, $K(z, \zeta)$ is a positive harmonic function of $\zeta \in R$ satisfying $K(z, \zeta_0) = 1$.

For the subsequent development of the integral representation (5.5), see Martin [1941] or Helms [1969], Chapter 12, for a modern treatment. Moreover, the normal family construction of $K(z, \zeta)$ is readily adapted to other settings; cf., e.g., Nakai [1966] for the existence of certain positive harmonic functions on subregions of a Riemann surface or Myrberg [1954] for the existence of positive solutions to the elliptic equation $\Delta u = Pu$.

5.5   Discontinuous Groups

In this section we turn our attention to the family $\mathcal{L}$ of *linear (fractional) transformations*

$$z' = T(z) = \frac{az+b}{cz+d}, \qquad ad - bc = 1, \ \ a, b, c, d \in \mathbb{C},$$

briefly considered in §3.8 from the standpoint of meromorphic functions. More general linear transformations for which $ad - bc$ is merely nonzero can be reduced to the above form by dividing each of the four constants by $\sqrt{ad - bc}$. Linear transformations are also known as *Möbius transformations, or bilinear transformations*, or, classically, *homographic substitutions*. The linear transformations map $\widehat{\mathbb{C}}$ one-to-one conformally onto itself, and form a group under composition. Indeed, the group $\mathcal{L}$ is isomorphic to the quotient group $\mathrm{SL}(2, \mathbb{C})/\{(I), (-I)\}$, where $(I) = \begin{pmatrix} 1 & 0 \\ 0 & 1 \end{pmatrix}$, and $\mathrm{SL}(2, \mathbb{C})$ is the *special linear group* of 2×2 matrices $\begin{pmatrix} a & b \\ c & d \end{pmatrix}$, with complex entries, having determinant one. For any $T, S \in \mathcal{L}$, the composition STS^{-1} is called the *conjugate* of T, and effects the transformation T, but in the new coordinate system defined by $w = S(z)$.

The study of discontinuous groups naturally arose out of work on elliptic modular functions and other automorphic functions. Important classical contributions to the field were made by Poincaré, Klein, and Schwarz, amongst others. The interested reader is referred to the comprehensive monograph of Lehner [1982].

Any linear transformation belonging to $\mathcal{L}$ can be classified according to the behaviour of its fixed points (or point). To this end, we set

$$z = T(z) = \frac{az+b}{cz+d},$$

and consider the resulting equation

$$cz^2 + (d-a)z - b = 0.$$

Case (1) : $c \neq 0$

This yields the two roots

$$\xi_1, \xi_2 = \frac{a - d \pm \sqrt{D}}{2c},$$

where $D = (d-a)^2 + 4bc = (a+d)^2 - 4$. Since $T(\infty) = a/c$, which is finite, there are at most two fixed points. On the other hand, if $D = 0$, that is, if the *trace* $a + d$ is ± 2 there is only one fixed point

$$\xi = \frac{a-d}{2c}.$$

Case (2) : $c = 0$

Then $a \neq 0$, $d \neq 0$, since $ad - bc = 1$. Since $T(z) = \frac{a}{d}z + \frac{b}{d}$, $\xi_1 = \infty$ is always a fixed point. Furthermore, if $a \neq d$, there is a second finite fixed point,

$$\xi_2 = \frac{b}{d-a}.$$

However, if $a = d \, (= \pm 1)$, then $T(z)$ is of the form

$$T(z) = z \pm b, \qquad (b \neq 0),$$

with the sole fixed point at ∞. Note that in this instance, $a + d = \pm 2$, so we find that in both Case (1) and Case (2), $T(z)$ has *exactly two fixed points if $D \neq 0$, and one fixed point if $D = 0$.*

Finally, the situation $c = 0$, $a = d$, $b = 0$, arises only from the identity transformation $I(z) = z$, $z \in \hat{\mathbb{C}}$, and thus we have shown that: *any linear transformation which is not the identity has at most two fixed points.*

As a consequence, *a linear transformation is uniquely determined by its values at three distinct points.* For suppose that $T(z_i) = z_i'$ and $S(z_i) = z_i'$, $i = 1, 2, 3$. Then $T \circ S^{-1}(z_i') = z_i'$, $i = 1, 2, 3$, so that by the preceding remarks, $T \circ S^{-1} = I$ and therefore $T = S$.

Moreover, given distinct points z_i, z_i', $i = 1, 2, 3$, lying in $\mathbb{C}$, the equation

$$\frac{(z' - z_1')(z_2' - z_3')}{(z' - z_2')(z_1' - z_3')} = \frac{(z - z_1)(z_2 - z_3)}{(z - z_2)(z_1 - z_3)} \tag{5.6}$$

engenders a linear transformation $T \in \mathcal{L}$ when z' is solved in terms of z, which satisfies $T(z_i) = z_i'$, $i = 1, 2, 3$. If one of z_1, or z_2, or z_3 is ∞, we replace the right-hand term of (5.6) with

$$-\frac{(z_2 - z_3)}{(z - z_2)}, \quad -\frac{(z - z_1)}{(z_1 - z_3)}, \quad \frac{(z - z_1)}{(z - z_2)},$$

respectively, by taking limits as the appropriate z_i becomes infinite. The left-hand term of (5.6) is altered similarly if any one of the z_i''s is ∞.

We are now in a position to classify the linear transformations of $\mathcal{L}$.

I.   One fixed point. We distinguish two cases, in both of which T is called *parabolic.*

(i) $c \neq 0$. Then the fixed point of $T(z)$ is given by $\xi = \frac{a-d}{2c}$, and $a+d = \pm 2$. Since the points $\infty, \xi, -d/c$ are mapped to $a/c, \xi, \infty$, respectively, the appropriate formulation of (5.6) gives

$$\frac{z' - \frac{a}{c}}{z' - \xi} = -\frac{\xi + \frac{d}{c}}{z - \xi}.$$

Subtracting the quantity 1 from both sides gives

$$\frac{\xi - \frac{a}{c}}{z' - \xi} = -\frac{\xi + \frac{d}{c}}{z - \xi} - 1.$$

Moreover,

$$\xi - \frac{a}{c} = \frac{a-d}{2c} - \frac{a}{c} = -\frac{a+d}{2c} = \mp\frac{1}{c}$$

and

$$\xi + \frac{d}{c} = \frac{a-d}{2c} + \frac{d}{c} = \frac{a+d}{2c} = \pm\frac{1}{c}.$$

Therefore, the transformation T reduces to the expression

$$\frac{1}{z' - \xi} = \frac{1}{z - \xi} \pm c, \tag{5.7}$$

where we have $+c$ if $a+d = 2$, and $-c$ if $a+d = -2$. By making the change of variable

$$w = S(z) = \frac{1}{z - \xi}, \quad w' = S(z') = \frac{1}{z' - \xi},$$

which map ξ to ∞, (5.7) can be written in the *normal form* for T,

$$w' = w \pm c,$$

where we note that $w' = STS^{-1}(w)$, with ∞ as the unique fixed point.

(ii) $c = 0$. Then $\xi = \infty$ is the unique fixed point, and as determined in Case (2), $a = d = \pm 1$, and

$$z' = z \pm b.$$

II. Two fixed points. Denote by ξ_1 and ξ_2 the two distinct fixed points of T. Again there are two cases.

(i) $c \neq 0$. The two fixed points of Case (1) are both finite, so that ξ_1, ξ_2, and ∞, are mapped to ξ_1, ξ_2, and a/c, respectively. Then the appropriate version of (5.6) has the form

$$\frac{(z' - \xi_1)\left(\frac{a}{c} - \xi_2\right)}{(z' - \xi_2)\left(\frac{a}{c} - \xi_1\right)} = \frac{(z - \xi_1)}{(z - \xi_2)},$$

which can be expressed as

$$\frac{z' - \xi_1}{z' - \xi_2} = k\frac{z - \xi_1}{z - \xi_2}, \tag{5.8}$$

where $k = \dfrac{a - c\xi_1}{a - c\xi_2}$. The number $k \neq 0, 1$ is the *multiplier* of T. Consider the transformations

$$w = S(z) = \frac{z - \xi_1}{z - \xi_2}, \quad w' = S(z') = \frac{z' - \xi_1}{z' - \xi_2},$$

which map the fixed points ξ_1, ξ_2 onto $0, \infty$, respectively, and satisfy $w' = STS^{-1}(w)$. Moreover, by (5.8),

$$w' = kw,$$

which is the *normal form* of T in this case, with $0, \infty$ as the fixed points.

(ii) $c = 0$. Then one of the fixed points is $\xi_1 = \frac{b}{d-a}$, and the other is $\xi_2 = \infty$, by Case (2). Hence

$$z' - \xi_1 = \frac{a}{d}z + \frac{b}{d} - \xi_1 = k(z - \xi_1),$$

where $k = \frac{a}{d}$. Taking $w = S(z) = z - \xi_1$, $w' = S(z') = z' - \xi_1$, which map ξ_1 and ∞, to 0 and ∞, we have, as above,

$$w' = kw.$$

The great virtue of writing the transformation T in terms of k is that when n applications of T are applied, namely T^n, it follows that

$$\frac{z' - \xi_1}{z' - \xi_2} = k^n \frac{z - \xi_1}{z - \xi_2} \quad \text{or} \quad z' - \xi_1 = k^n(z - \xi_1) \tag{5.9}$$

in (i) or (ii), respectively. Likewise, for the inverse transformation, we use k^{-1}, and for n applications, k^{-n}.

Furthermore, the classification of linear transformations with two fixed points is characterized by the multiplier $k = \rho e^{i\theta}$, $\rho > 0$, $0 \leq \theta < 2\pi$. If

$$\begin{aligned} k &= e^{i\theta}, \ \theta \neq 0, \text{ then } T \text{ is } \textit{elliptic}, \\ k &= \rho \neq 1, \text{ then } T \text{ is } \textit{hyperbolic}, \\ k &= \rho e^{i\theta}, \ \rho \neq 1, \ \theta \neq 0, \text{ then } T \text{ is } \textit{loxodromic}. \end{aligned}$$

The classification of linear transformations can also be formulated in terms of the trace $\chi = a + d$, in view of the relation (cf. Ford [1951], p. 16)

$$k + \frac{1}{k} = \chi^2 - 2,$$

which gives $k^{\frac{1}{2}} + k^{-\frac{1}{2}} = \chi$. Since a parabolic transformation has been shown to satisfy $\chi = \pm 2$, we can summarize our findings as: *A necessary and sufficient condition that $T \in \mathcal{L}$ ($T \neq I$) be elliptic, parabolic, or hyperbolic* *is that χ be real and $|\chi| < 2$, $|\chi| = 2$ or $|\chi| > 2$, respectively. T is loxodromic if and only if χ is complex.*

We note that for all $T, S \in \mathcal{L}$, both T and any conjugate STS^{-1} are of the same type (elliptic, etc.), since the multiplier and trace are invariant under conjugation.

Examples of Subgroups of $\mathcal{L}$.

(i) Simply Periodic Groups:

$$\{z' = z + n\omega : n \in \mathbb{Z}\}, \qquad \omega \neq 0.$$

(ii) Doubly Periodic Groups:

$$\{z' = z + n\omega + m\omega' : n, m \in \mathbb{Z}\}, \qquad \omega, \omega' \neq 0, \ \mathcal{I}m\left(\frac{\omega'}{\omega}\right) > 0.$$

(iii) Modular Group:

$$\mathsf{M} = \left\{z' = \frac{az + b}{cz + d} : ad - bc = 1, \ a, b, c, d \in \mathbb{Z}\right\}.$$

Elements of the modular group map the upper half-plane/lower half-plane/real axis onto themselves, as is seen from the fact that

$$\mathcal{I}m\,\frac{az + b}{cz + d} = \frac{y}{|cz + d|^2}, \qquad z = x + iy.$$

(iv) Picard Group:

$$\mathsf{P} = \left\{z' = \frac{az + b}{cz + d} : ad - bc = 1, \ a, b, c, d \text{ of the form } m + ni, \ m, n \in \mathbb{Z}\right\}.$$

The latter two groups are important examples in the sequel.

A few further preliminaries are necessary in order to establish the main results, the first of which is

Theorem 5.5.1 *Let D be an open disk in $\widehat{\mathbb{C}}$ and T a linear transformation such that $T(\overline{D}) \subset D$. Then T is either hyperbolic or loxodromic and $T(\overline{D})$ contains a fixed point of T.*

Proof. We may suppose that T is in normal form so that either $T(z) = z + \lambda$, or $T(z) = kz$, $\lambda \neq 0$, $k \neq 0, 1$. In the former case, $T(D)$ either overlaps D or is disjoint from it, excluding the possibility of T being parabolic. In the latter case, we may also exclude the possibility of T being elliptic, as this would simply result in a rotation of D about the origin. We conclude that T is either hyperbolic or loxodromic, having fixed points 0 and ∞.

If z is any value not fixed by T, then the iterations $T^n(z)$ converge to one of the fixed points of T (as $k \neq 0, 1$). Thus for $z \in D$, $T^n(z)$ must converge to a fixed point in $T(\overline{D})$, and the assertion is proved.

The next definition is central to our whole discussion.

Definition 5.5.2 *A group* Γ *of linear transformations is* **discontinuous at a point** α *if* α *is not in the closure of* $\Gamma(z) = \{T(z) : T \in \Gamma\}$, *for any* $z \in \widehat{\mathbb{C}}$. *In other words, there is no* $z \in \widehat{\mathbb{C}}$, *and no sequence of distinct elements* $T_n \in \Gamma$ *such that* $T_n(z) \to \alpha$ *as* $n \to \infty$. *In this instance,* α *is termed an* **ordinary point** *of* Γ; Γ *is* **discontinuous on a set** S *if it is discontinuous at each point of* S, *and* Γ *is a* **discontinuous group** *if it is discontinuous on some nonempty set.*

On the other hand, if there is a point z and an infinite sequence of distinct transformations $T_n \in \Gamma$ such that $T_n(z) \to \alpha$, then α is a *limit point* of Γ. Note that a point α which is a fixed point of infinitely many distinct $T_n \in \Gamma$ is necessarily a limit point.

The study of discontinuous groups has been inextricably linked to that of *automorphic functions*, that is, analytic functions $f(z)$ which take the same value at congruent points of a linear transformation group Γ:

$$f\big(T(z)\big) = f(z) \qquad \text{for all } T \in \Gamma.$$

If on a domain Ω, $T_n(z) \to \alpha \in \Omega$, then f automorphic in Ω means that it would be constant by the Identity Theorem. Hence there is the necessity for the groups associated with f to be discontinuous.

Let L (or $L(\Gamma)$) denote the set of limit points of a (discontinuous) group $\Gamma \subseteq \mathcal{L}$, and O (or $O(\Gamma)$) the set of ordinary points. Clearly, O is the complement of L.

The following properties require but routine verification.

Theorem 5.5.3 (i) *For each* $T \in \Gamma$, $T(L) = L$, $T(O) = O$.

(ii) *If* Δ *is a subgroup of* Γ, *then* $O(\Gamma) \subseteq O(\Delta)$.

(iii) $O(A\Gamma A^{-1}) = A\big(O(\Gamma)\big)$, $L(A\Gamma A^{-1}) = A\big(L(\Gamma)\big)$, *for any linear transformation* A.

From (ii) we see that any subgroup of a discontinuous group is discontinuous. Property (iii) allows one work with the transformed group and replace an ordinary point by ∞, say.

Discreteness. As noted previously, we may identify the transformation $z \to \frac{az+b}{cz+d}$ with the matrices $A = \big(\begin{smallmatrix} a & b \\ c & d \end{smallmatrix}\big)$ and $-A$. Then the discontinuity of a group Γ has a counterpart associated with the behaviour of the corresponding matrix group, which shall henceforth also be denoted by Γ.

We say that Γ is *discrete* if it contains no convergent sequence of distinct matrices $A_n \to A$. Here A need not belong to Γ and

$$A_n = \begin{pmatrix} a_n & b_n \\ c_n & d_n \end{pmatrix} \to A = \begin{pmatrix} a & b \\ c & d \end{pmatrix},$$

if and only if

$$\|A - A_n\| = \{|a - a_n|^2 + |b - b_n|^2 + |c - c_n|^2 + |d - d_n|^2\}^{\frac{1}{2}} \to 0$$

as $n \to \infty$; equivalently, $a_n \to a$, $b_n \to b$, $c_n \to c$, $d_n \to d$. Discreteness is tantamount to the identity matrix being isolated. Furthermore, the discreteness of Γ may be characterized by the condition:

$$\overline{\{A \in \Gamma : \|A\| \le n\}} < \infty, \qquad n = 1, 2, 3, \ldots . \tag{5.10}$$

For, on the one hand there can be no convergent sequence of distinct matrices if condition (5.10) is satisfied, implying Γ is discrete. On the other hand, if (5.10) does not hold, there must be infinitely many distinct $A_\nu \in \Gamma$ with $\|A_\nu\| \le n$, for some n. Therefore, a subsequence $\{A_{\nu_k}\}$ of $\{A_\nu\}$ converges to a matrix A. Since $a_{\nu_k} d_{\nu_k} - b_{\nu_k} c_{\nu_k} = 1$ for all ν_k, $A \in \mathrm{SL}(2, \mathbb{C})$, and Γ is not discrete.

As a consequence of the preceding characterization of discreteness we readily conclude

Theorem 5.5.4 *A discrete group Γ is countable.*

For,

$$\Gamma = \bigcup_{n=1}^{\infty} \Gamma_n,$$

where $\Gamma_n = \{A \in \Gamma : \|A\| \le n\}$, and each Γ_n is a finite set.

We further note that a conjugate group $B\Gamma B^{-1}$ of a discrete group Γ is itself discrete, since conjugation is a homeomorphism of $\mathrm{SL}(2, \mathbb{C})$.

Our first attempt to correlate discontinuity with discreteness is

Theorem 5.5.5 *A discontinuous group is discrete.*

Proof. Suppose that $A_n \to I$, for $A_n = \begin{pmatrix} a_n & b_n \\ c_n & d_n \end{pmatrix}$. Then for any $z \in \mathbb{C}$, we have $c_n z + d_n \neq 0$ for all n sufficiently large, and

$$\frac{a_n z + b_n}{c_n z + d_n} - z = \frac{-c_n z^2 + (a_n - d_n)z + b_n}{c_n z + d_n} \to 0$$

since $a_n \to 1$, $b_n \to 0$, $c_n \to 0$, $d_n \to 1$. This means that z is a limit point of the group which, as a consequence, is not discontinuous at z. Similarly for $z = \infty$.

A partial converse to this theorem will be established in Theorem 5.5.10.

One deduces from Theorems 5.5.4 and 5.5.5 that *a discontinuous group is countable*.

The Picard group P, on the other hand, affords an example of a discrete group which is not discontinuous. In fact, P is certainly discrete because of the discreteness of the Gaussian integers $m + ni$, $m, n \in \mathbb{Z}$. Moreover, for any given $z \in \widehat{\mathbb{C}}$, one can find relatively prime Gaussian integers α_n and γ_n such that $\alpha_n/\gamma_n \to z$. For α_n, γ_n thus determined, next find Gaussian integer solutions β_n, δ_n to $\alpha_n\delta_n - \beta_n\gamma_n = 1$. Then T_n given by $\left(\begin{smallmatrix} \alpha_n & \beta_n \\ \gamma_n & \delta_n \end{smallmatrix}\right)$ is a member of P with $T_n(\infty) = \alpha_n/\gamma_n \to z$, i.e., the Picard group is not discontinuous at any $z \in \widehat{\mathbb{C}}$.

Normality. In order to demonstrate that a family of linear transformations $\widehat{\mathcal{L}}$ is normal, it transpires that there is a somewhat stronger version of the FNT for meromorphic functions (§3.2), owing to the specific nature of $\widehat{\mathcal{L}}$. An arbitrary domain Ω in this section is assumed to lie on the Riemann sphere unless stated otherwise.

We shall first require some preliminary material which is of interest in its own right.

Lemma 5.5.6 *If $\widehat{\mathcal{L}} \subseteq \mathcal{L}$ is a normal family in Ω, then, for any $S \in \mathcal{L}$, the conjugate family $\{STS^{-1} : T \in \widehat{\mathcal{L}}\}$ is normal in $S(\Omega)$.*

The proof is readily verified and left to the reader.

As we shall be working with convergent sequences of linear transformations, it is necessary to know what form the limit may take. In this regard we have (cf. Piranian and Thron [1957]).

Theorem 5.5.7 *Suppose $\{T_n\} \subseteq \mathcal{L}$ and wherever the sequence converges define*

$$V(z) = \lim_{n \to \infty} T_n(z).$$

Whenever the range of V contains three distinct points, V is a linear transformation and convergence holds for all $z \in \widehat{\mathbb{C}}$. Furthermore, for suitable $\varepsilon_n = \pm 1$, $\varepsilon_n(T_n) \to (V)$ where $(T_n), (V)$ are matrices corresponding to T_n, V, respectively. If $T_n \to V$ uniformly on a compact region, then V is either a constant or a linear transformation such that $\varepsilon_n(T_n) \to (V)$.

Proof. It is evident that wherever $V(z)$ is defined, it is single-valued. Suppose that $V(z_i) = w_i$, $i = 1, 2, 3$, with the $\{w_i\}$ (and hence the $\{z_i\}$) being distinct. Next choose linear transformations R and S which satisfy

$$R(0) = z_1, R(1) = z_2, R(\infty) = z_3, S(w_1) = \infty, S(w_2) = -1, S(w_3) = 0,$$

and define $ST_nR = U_n$. Clearly $U_n(0) \to \infty$, $U_n(1) \to -1$, $U_n(\infty) \to 0$. From the latter limit one deduces that if U_n is given by $\left(\begin{smallmatrix} a_n & b_n \\ c_n & d_n \end{smallmatrix}\right)$, then

$$a_n/c_n \to 0. \tag{5.11}$$

Furthermore, writing

$$U_n(z) = \frac{a_n}{c_n} + \frac{1}{c_n^2}\frac{b_n c_n - a_n d_n}{z + d_n/c_n} = \frac{a_n}{c_n} - \frac{1}{c_n^2}\frac{1}{z + d_n/c_n},$$

we obtain from $U_n(0) \to \infty$ that

$$c_n d_n \to 0. \tag{5.12}$$

Finally, from $U_n(1) \to -1$, we also obtain

$$c_n(c_n + d_n) \to 1. \tag{5.13}$$

Combining (5.12) and (5.13) it follows that $c_n^2 \to 1$, and hence for a suitable sequence $\{\varepsilon_n\} = \{\pm 1\}$ we will have $\varepsilon_n c_n \to 1$. In view of (5.11), $a_n \to 0$, and, by (5.12), $d_n \to 0$. Moreover,

$$\varepsilon_n b_n = \frac{\varepsilon_n a_n \cdot \varepsilon_n d_n - 1}{\varepsilon_n c_n} \to -1,$$

by preceding limits.

One concludes that $\varepsilon_n(U_n) \to (U)$, where $(U) = \begin{pmatrix} 0 & -1 \\ 1 & 0 \end{pmatrix}$, and therefore $\varepsilon_n(T_n) = (S^{-1})\varepsilon_n(U_n)(R^{-1}) \to (S^{-1})(U)(R^{-1})$. As a consequence,

$$T_n(z) \to S^{-1}UR^{-1}(z) = V(z)$$

for all z, and $V(z)$ is a linear transformation.

Finally, on any compact region K, the univalence of T_n implies that V is either constant or univalent in K. If $V(z)$ is univalent it must assume infinitely many distinct values, and the preceding argument implies that V is a linear transformation and $\varepsilon_n(T_n) \to (V)$.

Corollary 5.5.8 *If Γ is a discrete group of linear transformations and $\{T_n\} \subseteq \Gamma$ is a sequence of distinct members satisfying $T_n \to \phi$ uniformly on a compact region, then ϕ is a constant.*

Proof. If ϕ is a linear transformation, then $\varepsilon_n(T_n) \to (\phi)$ by the theorem, contradicting the discreteness of Γ.

For some related results see Gehring and Martin [1987].

We are now able to establish a general criterion for normality (cf. Fatou [1930], §31, p. 71). A stronger version, of which the present result is a consequence, follows after Corollary 5.5.9.

FNT for Linear Transformations *Let $\widehat{\mathcal{L}}$ be a family of linear transformations defined on a domain Ω such that $T(z) \neq \omega_1$, $T(z) \neq \omega_2$, for all $z \in \Omega$, $T \in \widehat{\mathcal{L}}$, where ω_1, ω_2 are fixed complex numbers. Then $\widehat{\mathcal{L}}$ is normal in Ω.*

Proof. Given a sequence $\{T_\nu\} \subseteq \widehat{\mathcal{L}}$, the functions defined by

$$S_\nu(z) = \frac{T_\nu(z) - \omega_1}{T_\nu(z) - \omega_2}, \qquad z \in \Omega,$$

omit the values 0 and ∞ in Ω and are analytic there. Assume S_ν is given by $\begin{pmatrix} a_\nu & b_\nu \\ c_\nu & d_\nu \end{pmatrix}$, $\nu = 1, 2, 3, \ldots$.

If $\infty \notin \Omega$, it suffices to show that $\{S_\nu\}$ is normal in an arbitrary disk in Ω which we take to be the unit disk U. Since S_ν omits 0 and ∞, $|-b_\nu/a_\nu| > 1$, $|-d_\nu/c_\nu| > 1$. We now write

$$S_\nu(z) = \frac{b_\nu}{d_\nu} \frac{z + b_\nu/a_\nu}{b_\nu/a_\nu} \cdot \frac{d_\nu/c_\nu}{z + d_\nu/c_\nu} = \frac{b_\nu}{d_\nu} V_\nu(z),$$

which is certainly valid for $a_\nu c_\nu \neq 0$. Let us assume this to be the case for the present. Then for $|z| \leq r < 1$

$$0 < (1 - r)\frac{1}{1 + r} \leq |V_\nu(z)| \leq (1 + r)\frac{1}{1 - r} < \infty.$$

There are two cases to consider:

(i) If $|b_\nu/d_\nu| < M < \infty$, then $|S_\nu(z)| \leq M_1(r)$, and $\{S_\nu\}$ is normal, invoking Montel's theorem (§2.2).

(ii) There is a subsequence $|b_n/d_n| \to \infty$. In this instance, $|1/S_n(z)| \leq |d_n/b_n|M_2(r) \to 0$ uniformly in $|z| \leq r$. Then $S_n \to \infty$ uniformly in $|z| \leq r$, and the normality of $\{S_\nu\}$ is once again established.

In the event $a_j = 0$ for infinitely many indices j, then it must be that $c_j \neq 0$. In this case write

$$S_j(z) = \frac{b_j}{d_j} \frac{d_j/c_j}{z + d_j/c_j},$$

and the proof follows as before. Likewise if $c_j = 0$.

If $\infty \in \Omega$, we consider $S_\nu(1/z)$ and the result now follows.

Stated in terms of the contrapositive we obtain

Corollary 5.5.9 *If $\{T\}$ is not a normal family at a point α, then for all $w \in \mathbb{C}$, with at most one exception, w_0, there is a sequence of distinct T_n and a sequence $z_n \to \alpha$ for which $T_n(z_n) = w \neq w_0$.*

Proof. Let $\{D_i\}$ be a decreasing sequence of disks whose intersection is $\{\alpha\}$. As the family $\{T\}$ is not normal in D_1, $\cup_T T(D_1)$ omits at most one value, say w_0. Fix any $w \neq w_0$. Then there is some member T_1 of the family, and a point $z_1 \in D_1$ with $T_1(z_1) = w$. Since $\{T\} - \{T_1\}$ is not normal in

D_2 (but omits the value w_0), there is a member T_2 and a point $z_2 \in D_2$ satisfying $T_2(z_2) = w$ by the above reasoning. Continuing in this fashion establishes the result given an exceptional value w_0.

If there is no exceptional value, then the set of values $T(D_2)$ for $T \in \{T\}$ may omit some number w_1, but then must take every other value $w \neq w_1$. Then $T(D_2)$, for $T \in \{T\} - \{T_1\}$ cannot omit $w \neq w_1$, and so on.

There is a version of the FNT for Linear Transformations which is analogous to the Extended FNT for meromorphic functions (§4.1).

Extended FNT for Linear Transformations *Let $\widehat{\mathcal{L}}$ be a family of linear transformations on a domain Ω, and let $\varepsilon > 0$ be such that for any $T \in \widehat{\mathcal{L}}$ there are two points a_T, b_T omitted by T, satisfying*

$$\chi(a_T, b_T) > \varepsilon.$$

Then $\widehat{\mathcal{L}}$ is normal in Ω.

Before proceeding with the proof, it is necessary to discuss the following important notion. Let z_1, z_2, z_3, z_4 be four distinct points of $\mathbb{C}$, and define the *cross-ratio* to be

$$(z_1, z_2, z_3, z_4) = \frac{(z_1 - z_4)(z_2 - z_3)}{(z_2 - z_4)(z_1 - z_3)}.$$

(The cross-ratio was encountered implicitly in equation (5.6)). We extend this definition by continuity to the case when one of the z_i's is ∞, say, for example,

$$(z_1, z_2, z_3, \infty) = \frac{(z_2 - z_3)}{(z_1 - z_3)}.$$

The other cases follow similarly. Now, given a linear transformation

$$T(z) = \frac{az + b}{cz + d}$$

belonging to $\mathcal{L}$, we find, assuming $z, z', T(z), T(z') \in \mathbb{C}$, that

$$T(z) - T(z') = \frac{(z - z')}{(cz + d)(cz' + d)},$$

from which it follows that for any $z_i \in \widehat{\mathbb{C}}$, $T(z_i) \in \widehat{\mathbb{C}}$,

$$(z_1, z_2, z_3, z_4) = (T(z_1), T(z_2), T(z_3), T(z_4)).$$

In other words, *the cross-ratio is invariant under any $T \in \mathcal{L}$.* In view of the definition of the spherical metric χ (§1.2), one thereby deduces

$$\frac{\chi(z_1, z_4)\chi(z_2, z_3)}{\chi(z_2, z_4)\chi(z_1, z_3)} = \frac{\chi(T(z_1), T(z_4))\chi(T(z_2), T(z_3))}{\chi(T(z_2), T(z_4))\chi(T(z_1), T(z_3))}, \qquad z_i \in \widehat{\mathbb{C}}. \quad (5.14)$$

We may now proceed with the proof of the Extended FNT for Linear Transformations. Fix any $T \in \widehat{\mathcal{L}}$ and let z, w be distinct points of Ω, and α, β distinct points in the complement of Ω. Then the product of the cross-ratios $(z, \alpha, w, \beta) \cdot (z, \beta, w, \alpha)$ is invariant under T, and whence by (5.14),

$$\frac{\chi(z,\beta)\chi(\alpha,w)}{\chi(\alpha,\beta)\chi(z,w)} \cdot \frac{\chi(z,\alpha)\chi(\beta,w)}{\chi(\beta,\alpha)\chi(z,w)}$$

$$= \frac{\chi(T(z),T(\beta))\chi(T(\alpha),T(w))}{\chi(T(\alpha),T(\beta))\chi(T(z),T(w))} \cdot \frac{\chi(T(z),T(\alpha))\chi(T(\beta),T(w))}{\chi(T(\beta),T(\alpha))\chi(T(z),T(w))}.$$

As a consequence,

$$\left[\frac{\chi(T(z),T(w))}{\chi(z,w)}\right]^2$$

$$\leq \left[\frac{\chi(\alpha,\beta)}{\chi(T(\alpha),T(\beta))}\right]^2 \left[\frac{1}{\chi(z,\alpha)\chi(z,\beta)\chi(w,\alpha)\chi(w,\beta)}\right]$$

$$\leq \left[\frac{1}{\chi(T(\alpha),T(\beta))}\right]^2 \left[\frac{1}{\chi(z,\alpha)} + \frac{1}{\chi(z,\beta)}\right]\left[\frac{1}{\chi(w,\alpha)} + \frac{1}{\chi(w,\beta)}\right].$$

Writing, $\alpha = T^{-1}(a_T)$, $\beta = T^{-1}(b_T)$ gives

$$\chi(T(z),T(w)) \leq \frac{2\chi(z,w)}{\chi(a_T,b_T)\chi(z,\partial\Omega)^{\frac{1}{2}}\chi(w,\partial\Omega)^{\frac{1}{2}}}, \qquad T \in \widehat{\mathcal{L}}, \qquad (5.15)$$

which by the hypotheses implies that $\widehat{\mathcal{L}}$ is spherically equicontinuous in Ω. The normality of $\widehat{\mathcal{L}}$ can then be deduced as in the proof of Montel's theorem (§2.2), replacing the usual metric by χ, and taking into account the compactness of the Riemann sphere (compare Theorem 3.2.1). We remark that the constant 2 in the inequality (5.15) is best possible (cf. Beardon [1983], p. 43). Note that the hypotheses include the case of the family $\widehat{\mathcal{L}}$ omitting two fixed points.

As a culmination of much of the preceding work, a necessary and sufficient condition for a transformation group to be discontinuous can be stipulated in terms of normality (cf. Fatou [1930], §32, p. 74).

Theorem 5.5.10 *A group Γ of linear transformations is discontinuous at a point α if and only if Γ is discrete and forms a normal family at α.*

Proof. Let us assume Γ is infinite, for otherwise the result is trivial. First, suppose Γ is discontinuous at α. By Theorem 5.5.5, Γ is discrete. If $\Gamma = \{T_\nu\}$ is not normal at some point α, then there is a number w, transformations T_n, and a sequence $z_n \to \alpha$ such that $T_n(z_n) = w$ (Corollary 5.5.9). Therefore, $z_n = T_n^{-1}(w) \to \alpha$, but this contradicts the discontinuity of Γ at α.

Conversely, suppose that Γ is discrete and normal at α, but *not* discontinuous at α. In view of Lemma 5.5.6, we may assume α to be finite. Then there is a point z_0 and distinct $T_\nu \in \Gamma$ such that $T_\nu(z_0) \to \alpha$. Let D be an open disk centred at α in which Γ is normal, and choose T_N satisfying $T_N(z_0) = z_1 \in D$. Next, define

$$S_\nu = T_\nu \circ T_N^{-1},$$

so that $S_\nu(z_1) \to \alpha$. Now $\{S_\nu\} \subseteq \Gamma$ and hence is normal in D. Then there is a subsequence $\{S_\mu\}$ converging uniformly on a compact subdisk $\overline{D}_1 \subseteq D$, D_1 centred at α, $z_1 \in \overline{D}_1$.

According to Corollary 5.5.8, $\{S_\mu\}$ converges to a constant, which in this case must be α. Thus, for μ_0 sufficiently large, $S_{\mu_0}(\overline{D}_1)$ is a proper subset of D_1. This means that S_{μ_0} must be either hyperbolic or loxodromic (Theorem 5.5.1), and that $S_{\mu_0}(\overline{D}_1)$ contains a fixed point β of S_{μ_0}. Hence, by (5.9), $S_{\mu_0}^n$ can be expressed in the form

$$\frac{z'-\beta}{z'-\gamma} = k^n \frac{z-\beta}{z-\gamma}, \qquad |k| \neq 1, \ \gamma \text{ finite,}$$

or

$$z' - \beta = k^n(z - \beta), \qquad |k| \neq 1, \ \gamma = \infty.$$

In either case, observe that $S_{\mu_0}^n(\beta) = \beta$ for all n. On the other hand, for any $z \neq \beta$, $S_{\mu_0}^n(z) \to \gamma$ as $n \to \infty$ ($|k| > 1$) or as $n \to -\infty$ ($|k| < 1$). As a consequence, $\{S_{\mu_0}^n(z)\}_{n\in\mathbb{Z}}$ cannot be normal in D_1, and hence the larger family Γ cannot be normal there either. This contradiction proves the result.

From the preceding theorem we may deduce the following formulation.

Theorem 5.5.11 *Let Γ be a discrete group of linear transformations. Then Γ is discontinuous at each point z which is covered by a disk D such that $\Gamma(D)$ omits at least two points of $\widehat{\mathbb{C}}$. Furthermore, Γ is discontinuous on an open set S if $\Gamma(S) \subseteq S$ and S omits two points.*

Thus the modular group M, which is obviously discrete, is discontinuous throughout $\mathcal{H}_+$ and $\mathcal{H}_-$.

In the event $\Gamma(D)$ covers $\widehat{\mathbb{C}}$, as is the case for the Picard group, then we have seen that the result no longer holds. The whole matter of when a discrete group is discontinuous has been taken a step further with a theorem of Larcher [1963]: *A discrete group of linear transformations Γ is discontinuous if and only if there is a disk D such that $\Gamma(D)$ omits at least one point of $\widehat{\mathbb{C}}$.* Part of Larcher's argument invokes the foregoing two-point criterion.

As consequence of this result we have

Corollary 5.5.12 *If all the members of a discrete group Γ are analytic in a domain Ω, then Γ is discontinuous.*

Corollary 5.5.13 *If Γ is a discrete group of automorphisms of a domain $\Omega \neq \widehat{\mathbb{C}}$, then Γ is discontinuous.*

There is another notion in the literature (cf. Ford [1951], p. 35, Behnke and Sommer [1965], p. 513) connected with a transformation group being discontinuous, and that is the notion of being *properly discontinuous*. Indeed, the two concepts are equivalent.

Definition 5.5.14 *A group of linear transformations Γ is called **properly discontinuous** if there is a point α and a neighbourhood N_α of α such that for each $T \in \Gamma$, $T \neq I$, $T(\alpha) \notin N_\alpha$. A point possessing this property is referred to as a **standard point**.*

Theorem 5.5.15 *A group of linear transformations Γ is discontinuous if and only if it is properly discontinuous.*

Proof. First suppose that Γ is discontinuous, and let α be an ordinary point which is not fixed by any member of Γ. As there are uncountably many ordinary points (which follows from Theorem 5.5.10) and only countably many transformations, each with at most two fixed points, such a point α surely exists. Since $\Gamma(\alpha)$ does not accumulate at α and $T(\alpha) \neq \alpha$, for $T \in \Gamma$, $T \neq I$, it follows that α is a standard point, that is, Γ is properly discontinuous.

Conversely, assume Γ is properly discontinuous, and that the standard point is $\alpha = \infty$. We prove that Γ is discrete, and normal at ∞. If Γ is not discrete, then there is a sequence of distinct $T_\mu \in \Gamma$ satisfying $T_\mu(z) \to z$ for all z. Then $T_\mu(\infty) \to \infty$ contradicting the fact that ∞ is a standard point.

To prove the normality of Γ at ∞, observe that the hypothesis implies that for any sequence $\{T_\nu\} \subseteq \Gamma$, $|T_\nu(\infty)| = |a_\nu/c_\nu| < M$, $|T_\nu^{-1}(\infty)| = |d_\nu/c_\nu| < M$, for some M. In a neighbourhood $N = \{|z| > 2M\}$, we have for

$$T_\nu(z) = \frac{a_\nu}{c_\nu} + \frac{1}{c_\nu^2}\frac{b_\nu c_\nu - a_\nu d_\nu}{z + d_\nu/c_\nu}$$

that

$$|T_\nu(z)| \leq \left|\frac{a_\nu}{c_\nu}\right| + \left|\frac{1}{c_\nu^2}\right|\frac{1}{|z + d_\nu/c_\nu|} \leq M + \frac{1}{|c_\nu^2|M}.$$

Since Γ is properly discontinuous, it is an elementary fact that $|c_\nu| \to \infty$ (cf. Ford [1951], pp. 40–41). Therefore, $|T_\nu(z)| \leq M_1$ for all $z \in N$, i.e., Γ is bounded in a neighbourhood of infinity, and the normality follows.

Appendix

Quasi-Normal Families

This notion was introduced by Montel [1922], and while not as far-reaching as the concept of normal families, has some worthwhile consequences. We develop only the most salient features, although much of the normal family theory can be generalized to this setting. Refer to Montel [1924, 1927], Valiron [1929], for further details, or the very informative *Commentary* by Valiron in *Selecta* [1947].

Definition A.1 *Let $\mathcal{F}$ be a family of analytic (meromorphic) functions on a domain Ω. Then $\mathcal{F}$ is **quasi-normal** on Ω if every sequence of functions $\{f_n\} \subseteq \mathcal{F}$ contains a subsequence which converges uniformly (spherically uniformly) on compact subsets of $\Omega - Q$, where Q is a (possibly empty) finite set of points of Ω. This set Q of exceptional points may vary with the particular sequence and constitutes the set of **irregular points**.*

The convergence is either to an analytic (meromorphic) function or to the function $\equiv \infty$ on $\Omega - Q$. If the set of irregular points never surpasses q in number, yet for some sequence there are q such points, then q is the *order* of the quasi-normal family. We first treat quasi-normal families of analytic functions.

Proposition A.2 *Let $\mathcal{F}$ be a quasi-normal family of analytic functions on Ω and suppose that a sequence $\{f_n\} \subseteq \mathcal{F}$ contains a subsequence $\{f_{n_k}\}$ which converges uniformly on every compact subset of $\Omega - Q$, but not on any compact subset containing points of Q, where Q is a nonempty finite set. Then the limit function of the subsequence must be $\equiv \infty$ on $\Omega - Q$.*

Proof. Suppose that the limit function $f(z)$ is in fact analytic on $\Omega - Q$. Choose $z_0 \in Q$ and take a small circle $C_r : |z - z_0| = r$ in Ω, where C_r contains no other points of Q. Since $f_{n_k} \to f$ uniformly on C_r, we can deduce from the Weierstrass Theorem (§1.4) that $f_{n_k} \to f$ uniformly on $K(z_0; \frac{r}{2})$, a contradiction. Thus f must be $\equiv \infty$.

Example A.3 Define $\mathcal{F} = \{\lambda z : \lambda \in \mathbb{C}\}$ for $z \in U$. If $\{\lambda_n\}$ is a sequence with $\lambda_n \to \lambda_0$, then $\lambda_n z \to \lambda_0 z$ uniformly in U, that is, in this instance

$Q = \emptyset$. However, if $\lambda_n = n$, then $\{nz\}$ converges to ∞ uniformly on compact subsets of $U - \{0\}$. Whence $\mathcal{F}$ is quasi-normal of order 1, with the origin as the sole irregular point.

More generally,

$$\mathcal{F} = \{\lambda P(z) : P(z) \text{ a polynomial of degree } q, \ \lambda \in \mathbb{C}\}$$

is quasi-normal of order q, with q irregular points at the zeros of $P(z)$.

In order to prove an analogue of the FNT, we further require

Proposition A.4 *Let $\{f_n\}$ be a sequence in a quasi-normal family $\mathcal{F}$ on Ω. Suppose that for some fixed $z_0 \in \Omega$, no subsequence of $\{f_n\}$ converges uniformly on any compact subset of Ω containing z_0. Then, if $a \in \mathbb{C}$ and N is a fixed neighbourhood of z_0, each of the equations $f_n(z) = a$ has a root in N, for all sufficiently large n.*

Proof. Assume that the conclusion does not hold. Then there is a number $a \in \mathbb{C}$ and a neighbourhood $N = K(z_0; r) \subseteq \Omega$, as well as a subsequence $\{\phi_n\} \subseteq \{f_n\}$ such that $\phi_n(z) = a$ has no root in N. From the sequence $\{\phi_n\}$ we can extract a subsequence $\{\phi_{n_k}\}$ which converges uniformly on compact subsets of $\Omega - Q$, for some finite set Q of irregular points. Since $z_0 \in Q$, Proposition A.2 implies $\phi_{n_k} \to \infty$ uniformly on compact subsets of $\Omega - Q$, in particular on the circle $C_r : |z - z_0| = r$. Then the function

$$\psi_k(z) = \frac{1}{\phi_{n_k}(z) - a}$$

is analytic in $D(z_0; r)$ and tends to zero uniformly on C_r. By the Maximum Modulus Principle, $\psi_k \to 0$ uniformly in $D(z_0; r)$, i.e., $\phi_{n_k} \to \infty$ uniformly in $D(z_0; r)$, contradicting the hypotheses.

For a family to be quasi-normal, there is the following version of the Fundamental Normality Test (§2.7).

Theorem A.5 *Let $\mathcal{F}$ be a family of analytic functions on a domain Ω which do not take a value $a \in \mathbb{C}$ more than q times, nor a value $b \in \mathbb{C}$ more than p times. Then $\mathcal{F}$ is quasi-normal of order $\leq \min\{p, q\}$.*

Proof. Without loss of generality, assume that $a = 0$, $b = 1$. Choose a sequence $\{f_n\} \subseteq \mathcal{F}$, and let z_1 in Ω be a point of accumulation of the zeros of the f_n's if such a point interior to Ω exists. From $\{f_n\}$ extract a subsequence $\{f_{\alpha_n}\}$, each function of which has a zero in some fixed arbitrarily small closed disk Δ_1 centred about z_1. Suppose that there is in Ω a second point of accumulation z_2 of the zeros of the functions $\{f_{\alpha_n}\}$. We may assume that z_2 is exterior to Δ_1, since removing if needs be a finite number of terms at the beginning of the sequence $\{f_{\alpha_n}\}$, we may take the radius of Δ_1 as small as we like. Place a closed disk $\Delta_2 \subseteq \Omega$ about z_2 exterior to Δ_1. From $\{f_{\alpha_n}\}$ extract a subsequence $\{f_{\beta_n}\}$ each of whose functions has at least one zero in Δ_2. We may have to go on in this manner finding $q' \le q$ points $z_1, z_2, \ldots, z_{q'}$, and a subsequence $\{f_{\lambda_n}\}$ whose functions each have a zero in the disks $\Delta_1, \Delta_2, \ldots, \Delta_{q'}$.

Next, let Ω' be a relatively compact subdomain with $\overline{\Omega'} \subseteq \Omega$ and $\cup_{i=1}^{q'}\Delta_i \subseteq \Omega'$. Set $\Omega'' = \Omega' - \cup_{i=1}^{q'}\Delta_i$. Then $\{f_{\lambda_n}\}$ has only a finite number of zeros in Ω'', so that for sufficiently large n, $\{f_{\lambda_n}\}$ has no zeros in Ω''.

Starting with $\{f_{\lambda_n}\}$, repeat the above procedure with respect to the roots of the equation

$$f_{\lambda_n}(z) = 1.$$

We obtain $p' \le p$ points $\zeta_1, \zeta_2, \ldots, \zeta_{p'}$ (some of which may coincide with the points z_i) and a subsequence $\{f_{\mu_n}\}$, each of whose functions has a root of $f_{\mu_n}(z) = 1$ in a small neighbourhood about the points $\zeta_1, \zeta_2, \ldots, \zeta_{p'}$. Since $\{f_{\mu_n}\}$ may be assumed to have no zeros in Ω'', and each function takes the value 1 at most p times, the Generalized Normality Test (§2.7) implies there is a subsequence $\{f_{\nu_n}\}$ converging uniformly on compact subsets of Ω''.

There are two cases to consider.

(i) The limit function is analytic in Ω''. As in the proof of Proposition A.2, we deduce that $\{f_{\nu_n}\}$ converges to an analytic function on each of the disks Δ_i. Therefore $\{f_{\nu_n}\}$ converges to an analytic function on compact subsets of Ω' and likewise on compact subsets of Ω.

(ii) The limit function is identically ∞ on Ω''. Then $\{f_{\nu_n}\}$ indeed converges uniformly to ∞ on any compact subset of $\Omega - \{z_1, \ldots, z_{q'}\}$. This is a consequence of the fact that we can make the disk Δ_i surrounding each z_i arbitrarily small, by going sufficiently far out in the sequence $\{f_{\nu_n}\}$ for the desired convergence.

As no irregular points arise from (i), we need only consider (ii) further. We claim that the only irregular points come from the points in common between the z_i and ζ_j, that is, if z_i is distinct from all the ζ_j, then z_i is not an irregular point. Indeed, suppose that no subsequence of $\{f_{\nu_n}\}$ converges uniformly on any compact subset of Ω containing z_i (i.e., z_i is an irregular point). Mimicking the proof of Proposition A.4 we can show that for all sufficiently large n, $f_{\nu_n}(z) = 1$ has a root in a neighbourhood of z_i. However, if z_i is distinct from all the ζ_j there can be no such root for all sufficiently large n. This proves the assertion. We conclude that $\mathcal{F}$ is quasi-normal and any irregular point must be one in common with some z_i and ζ_j, so that the order is at most $\min\{p, q\}$.

Other criteria for a family to be quasi-normal can be found in the thesis of Marty [1931] and in Montel [1936, 1939]. A particular class of irregular points, namely those of infinite order, has been studied by Saxer [1930]. These are engendered by the Proposition A.4: An irregular point ζ is an *irregular point of order* μ if there exists an infinite number of functions $f(z)$

of the family such that $f(z) - a$ has, for every given $a \in \mathbb{C}$, at least μ zeros in a neighbourhood of ζ, and if there is only a finite number of functions $f(z)$ for which $f(z) - a$ has more than μ zeros in the neighbourhood of ζ. In the contrary case, ζ has *infinite order*. For other work on irregular points, see Ostrowski [1926] and Leau [1932], Section 6.

It is often useful to know when a quasi-normal family is actually normal. To this end we have

Theorem A.6 *Let $\mathcal{F}$ be a quasi-normal family of anlytic functions in a domain Ω of order at most q. If the functions of $\mathcal{F}$ are bounded at $q + 1$ points, then $\mathcal{F}$ is normal.*

Proof. Consider a sequence $\{f_n\} \subseteq \mathcal{F}$. We can find a subsequence $\{f_{n_k}\}$ which converges uniformly to an analytic function, or coverges uniformly to ∞, on compact subsets which omit at most q irregular points. However, by assumption there must be at least one nonirregular point at which all the f_{n_k} are bounded, that is, there can be no irregular points with respect to $\{f_{n_k}\}$ in view of Proposition A.2, and so $\mathcal{F}$ is normal.

Furthermore, in this regard there is

Theorem A.7 *Suppose that $\mathcal{F}$ is a quasi-normal family of analytic functions in Ω, each of which does not take a fixed value $\omega \in \mathbb{C}$ more than q times. If the values of each $f \in \mathcal{F}$ and the first q derivatives are all bounded at some one point in Ω, then $\mathcal{F}$ is a normal family.*

Proof. Without loss of generality we may take $\omega = 0$, and the point in Ω where f and its first q derivatives are bounded as the origin. As in the preceding proof, we show that there can be no irregular points by demonstrating that no sequence of functions $\{f_n\}$ can tend to ∞ on the complement of a finite set of points.

In fact, suppose that $\{f_n\}$ converges to ∞ uniformly on compact subsets which omit the irregular points. Let C be an arbitrarily small circle centred about the origin and consider the Taylor expansion for each f_n

$$
\begin{aligned}
f_n(z) &= a_0^{(n)} + a_1^{(n)} z + \ldots + a_q^{(n)} z^q + a_{q+1}^{(n)} z^{q+1} + \ldots \\
&= P_n(z) + g_n(z),
\end{aligned}
$$

where $P_n(z) = a_0^{(n)} + a_1^{(n)} z + \ldots + a_q^{(n)} z^q$, $g_n(z) = a_{q+1}^{(n)} z^{q+1} + \ldots$.

By hypothesis, the coefficients $a_0^{(n)}, \ldots, a_q^{(n)}$ are bounded for each n, that is, $|P_n(z)| < M$, $z \in C$, all n. Moreover, taking n sufficiently large, $z \in C$, we also have $|f_n(z)| > M$, i.e., $|P_n(z)| < |f_n(z)|$, $z \in C$. By Rouché's theorem, f_n and g_n have the same number of zeros inside C. But $g_n(z) = a_{q+1}^{(n)} z^{q+1} + \ldots$ has at least $q + 1$ such zeros, whereas f_n has at most q, and this contradiction yields the normality of $\mathcal{F}$.

Corollary A.8 *Consider the functions whose first $q+1$ terms are given by*

$$f(z) = a_0 + a_1 z + \ldots + a_q z^q + \ldots + a_n z^n + \ldots,$$

which are analytic in $|z| < R$ and which take the value 0 no more than q times, nor the value 1 more than p times. Then, for $0 < \theta < 1$,

$$|f(z)| \le M, \qquad |z| \le \theta R,$$

where $M = M(\theta, a_0, a_1, \ldots, a_q)$.

In fact, by Theorem A.5 the family of such functions is quasi-normal, and by the preceding theorem it is normal. The result then follows as in the proof of Schottky's theorem (§2.8), which is the special case $p = q = 0$. A similar result was proved by Bieberbach [1922].

Other applications of the theory of quasi-normal families can be found in the works of Montel and Valiron mentioned earlier, as well as in Saxer [1928] concerning Picard's theorems, Cartwright [1935b] on precise inequalities in function theory, and Montel [1937] in a study of local valence.

Meromorphic Families. In discussing quasi-normal families of meromorphic functions, a significant difference arises in contrast with the analytic case regarding the conclusion of Proposition A.2. In the present circumstance, the set Q of irregular points may be nonvoid, yet a subsequence may converge to a meromorphic function on compact subsets exterior to Q.

For example, consider the sequence of functions

$$f_n(z) = f(z) + \frac{g(z) - f(z)}{1 + nP(z)},$$

where $f(z)$ and $g(z)$ are two distinct analytic functions and $P(z)$ a polynomial. Then $f_n \to f$ uniformly on compact subsets exterior to the roots of $P(z)$, whereas $f_n \to g$ at the roots of $P(z)$.

In spite of this difference, there is a meromorphic version of Theorem A.5 which is proved in a similar fashion (Montel [1927], p. 149).

Theorem A.9 *Let $\mathcal{F}$ be a family of meromorphic functions on a domain Ω which do not take a value a more than p times, a value b more than q times, nor a value c more than r times, with $p \ge q \ge r$. Then $\mathcal{F}$ is quasi-normal of order at most q.*

Investigations into families of meromorphic functions (termed *hyponormal*) having a set of irregular points which is no longer finite, but of zero linear measure, have been carried out by Bloch [1926c].

References

The numbers in brackets refer to the pages in the text where the work in question is cited.

Ahlfors, L.V.
Beiträge zur Theorie der meromorphen Funktionen, *C.R. 7ᵉ Congr. Math. Scand.* Oslo (1929), 84–88. [26]
Zur Theorie der Überlagerungsflächen, *Acta Math.* **65** (1935), 157–194. [29, 58]
An extension of Schwarz' Lemma, *Trans. Amer. Math. Soc.* **43** (1938), 359–364. [58]
Complex Analysis, 3rd ed., McGraw-Hill, New York, 1979. [1, 14, 18, 20, 44, 191]

Anderson, J.M., Clunie, J., Pommerenke, Ch.
On Bloch functions and normal functions, *J. Reine Angew. Math.* **270** (1974), 12–37. [182]

Anderson, J.M., Rubel, L.A.
Hypernormal meromorphic functions, *Houston J. Math.* (3) **4** (1978), 301–309. [181]

Arzelà, C.
Funzioni di linee, *Rend. della R. Accad. Lincei* **4** (1889), 342–348. [35]
Sulle funzioni di linee, *Mem. Accad. Sci. Bologna* (5) **5** (1895), 225–244. [35]
Sul principio di Dirichlet, *Rend. Accad. Sci. Bologna nouv. serie* **1** (1896–1897), 71–84. [viii]
Sulle serie di funzioni, I., *Mem. della R. Accad. Bologna* (5) **8** (1899), 3–58. II. ibid. 91–134. [35]

Ascoli, G.
Le curve limiti di una varietà data di curve, *Mem. della R. Accad. Lincei* **18** (1883), 521–586. [14, 35]

Bagemihl, F.
Some identity and uniqueness theorems for normal meromorphic functions, *Ann. Acad. Sci. Fenn. Ser. A.I.* no. 299, 1961. [182]
Some boundary properties of normal functions bounded on nontangential arcs, *Arch. Math.* **14** (1963), 399–406. [182]

Bagemihl, F., Seidel, W.
Behavior of meromorphic functions on boundary paths, with applications to normal functions, *Arch. Math.* **11** (1960), 263–269. [182]
Koebe arcs and Fatou points of normal functions, *Comm. Math. Helv.* **36** (1961), 9–18. [182]

Baker, I.N.
Repulsive fixed points of entire functions, *Math. Zeit.* **104** (1968), 252–256. [viii, 177]
Limit functions and sets of non-normality in iteration theory, *Ann. Acad. Sci. Fenn.* No. 467 (1970), 1–11. [178]

Beardon, A.F.
Montel's theorem for subharmonic functions and solutions of partial differential equations, *Proc. Camb. Phil. Soc.* **69** (1971), 123–150. [186, 192]
The Geometry of Discrete Groups, Springer-Verlag, New York, 1983. [6, 205]
Iteration of Rational Functions – Complex Analytic Dynamical Systems, Springer-Verlag, New York, 1991. [167, 175, 177]

Behnke, H., Sommer, F.
Theorie der analytischen Funktionen einer komplexen Veränderlichen, Springer-Verlag, Berlin, 1965. [1, 207]

Bermant, A.F.
Dilation d'une fonction modulaire et problèmes de recouvrement, *Mat. Sb.* **15** (57) (1944), 285–324. [67]

Bermant, A.F., Lavrentieff, M.A.
Sur l'ensemble de valeurs d'une fonction analytique, *Mat. Sb.* **42** (1935), 435–450. [65]

Bieberbach, L.
Über die Verteilung der Null- und Einsstellen analytischen Funktionen, *Math. Ann.* **85** (1922), 141–148. [213]

Biernacki, M.
Sur les fonctions univalentes, *Mathematica* **12** (1936), 49–64. [53]

Blanchard, P.
Complex analytic dynamics on the Riemann sphere, *Bull. Amer. Math. Soc.* **11** (1984), 85–141. [167, 173, 175]

Blaschke, W.
Eine Erweiterung des Satzes von Vitali über Folgen analytischer Funktionen, *Ber. Verhandl. Kön. Sächs. Gesell. Wiss. Leipzig* **67** (1915), 194–200. [47]

Bloch, A.
Démonstration directe de théorèmes de M. Picard, *C.R. Acad. Sci. Paris* **178** (1924), 1593. [56]
Les théorèmes de M. Valiron sur les fonctions entières et la théorie de l'uniformisation, *Ann. Fac. Sci. Univ. Toulouse* (1925), 1–22. [112]
(a) *Les fonctions holomorphes et méromorphes dans le cercle unité*, Gauthier-Villars, Paris, 1926. [105]
(b) La conception actuelle de la théorie des fonctions entières et méromorphes, *L'Enseignement Math.* 25 (1926), 83–103. [30, 101]
(c) Sur les systèmes de fonctions à variétés linéaires lacunaires, *Ann. École Norm. Sup.* (3) **43** (1926), 309–362. [213]

Bohr, H.
On the limit values of analytic functions, *J. London Math. Soc.* **2** (1927), 180–181. [64]

Bonk, M.
On Bloch's constant, *Proc. Amer. Math. Soc.* **110** (4) (1990), 889–894. [112]

Borel, E.
Démonstration élémentaire d'un théorème de M. Picard sur les fonctions entières, *C.R. Acad. Sci. Paris* **72** (1896), 1045–1048. [56]

Bowen, N.A.
On the limit of the modulus of a bounded regular function, *Proc. Edin. Math. Soc.* (2) **14** (1964–65), 21–24. [64]

Brolin, H.
Invariant sets under iteration of rational functions, *Arkiv För Mat.* **6** (1965), 103–144. [176]

Burckel, R.B.
An Introduction to Classical Complex Analysis, Vol. 1, Birkhäuser Verlag, Basel, 1979. [48, 58, 63, 64]

Bureau, F.
Mémoire sur les fonctions uniformes à point singulier essentiel isolé, *Mém. Soc. Roy. Sci. Liége* (3) **17** (1932). [101, 117]

Carathéodory, C.
Sur quelques applications du théoréme de Landau-Picard, *C. R. Acad. Sci. Paris* **144** (1907), 1203–1206. [67]
Untersuchungen über die konformen Abbildungen von festen und veränderlichen Gebieten, *Math. Ann.* **72** (1912), 107–144. [67]
Stetige Konvergenz und normale Familien von Funktionen, *Math. Ann.* **101** (1929), 515–533. [74]
Theory of Functions of a Complex Variable, Vol. I and II, Chelsea Publ. Co., New York, 1958 and 1960. [46, 58, 59, 74, 79, 97, 104]

Carathéodory C., Landau, E.
Beiträge zur Konvergenz von Funktionenfolgen, *Sitz. Kön. Preuss. Akad. Wiss. Berlin* (1911), 587–613. [47, 54]

Cartan, H.
Sur les systèmes de fonctions holomorphes à variétés linéaires lacunaires et leurs applications, *Ann. École Norm. Sup.* (3) **45** (1928), 255–346. [101]
Sur la fonction de croissance attachée à une fonction méromorphe de deux variables, et ses application aux fonctions méromorphes d'une variable, *C.R. Acad. Sci. Paris* **189** (1929), 521-523. [25]
Elementary Theory of Analytic Functions of One or Several Complex Variables, Addison-Wesley Co., Reading, MA, 1963. [8]

Cartwright, M.L.
(a)   Some generalizations of Montel's theorem, *Proc. Camb. Phil. Soc.* **31** (1935), 26–30. [63]
(b)   Some inequalities in the theory of functions, *Math. Ann.* **111** (1935), 98–118. [213]
A generalization of Montel's theorem, *J. London Math. Soc.* **37** (1962), 179–184. [64]

Chen, H., Gu, Y. (Ku, Y.)
An improvement of Marty's criterion and its applications, preprint (1992). [77, 103, 137]

Chen, H., Hua , X.
Normal families of holomorphic functions, preprint (1992). [137]

Chuang, C.T.
Sur les fonctions holomorphes dans le cercle unité, *Bull. Soc. Math. France* **68** (1940), 11–41. [125]

Clunie, J.
> On a result of Hayman, *J. London Math. Soc.* **42** (1967), 389–392.
> [132, 144]

Coddington, E.A., Levinson, N.
> *Theory of Ordinary Differential Equations*, McGraw-Hill, New York,
> 1955. [128]

Collingwood, E.F., Lohwater, A.J.
> *The Theory of Cluster Sets*, Cambridge University Press, Cambridge,
> 1966. [179]

Courant, R.
> *Dirichlet's Principle, Conformal Mapping and Minimal Surfaces*, In-
> terscience Publ. Inc., New York, 1950. [162]

de Branges, L.
> A proof of the Bieberbach conjecture, *Acta Math.* **154** (1985), 137–
> 152. [163]

de la Vallée Poussin, C.
> Démonstration simplifiée du théorème fondamental sur les familles
> normales de fonctions, *Ann. Math.* **17** (2) (1915), 5–11. [55]

de Possel, R.
> Zum Parallelschlitztheorem unendlich-vielfach zusammenhängender
> Gebiete, *Göttingen Nachr.* (1931), 199–202. [53, 163]

Devaney, R.L.
> Julia sets and bifurcation diagrams for exponential maps, *Bull. Amer.*
> *Math. Soc.* **11** (1984), 167–171. [178]
> *An Introduction to Chaotic Dynamical Systems*, Addison-Wesley Co.,
> Reading, MA, 1987. [167]

Drasin, D.
> Normal families and the Nevanlinna theory, *Acta Math.* **122** (1969),
> 231–263. [vii, 87, 113, 117, 123, 130, 137, 142]
> The impact of Lars Ahlfors's work in value distribution theory, *Ann.*
> *Acad. Sci. Fenn. Ser. A.I.* (3) **13** (1988), 329–353. [87]

Dufresnoy, J.
> Sur les domaines couvertes par les valeurs d'une fonction méromorphe
> ou algebroïde, *Ann. École Norm. Sup.* (3) **58** (1941), 179–259. [83]

Duren, P.L.
> *Theory of H^p Spaces*, Academic Press, New York, 1970. [63]
> *Univalent Functions*, Springer-Verlag, New York, 1983. [164, 167]

Fatou, P.
   Sur les équations fonctionelles, *Bull. Soc. Math. France* **47** (1919),
   161–271; **48** (1920), 33–94 and 208–314. [viii, 167]
   Sur l' itération des fonctions transcendantes entières, *Acta Math.* **47**
   (1926), 337–370. [177, 178]
   *Théorie des fonctions algébriques d'une variable et des transcendantes
   qui s' y rattachent. Tome II, Fonctions automorphes*, Gauthier-Villars,
   Paris, 1930. [202, 205]

Fekete, M.
   Zum Koebeschen Verzerrungssatz, *Göttingen Nachr.* (1925), 142–150.
   [67]

Ferrand, J.
   Étude de la représentation conforme au voisinage de la frontière, *Ann.
   École Norm. Sup.* (3) **59** (1942), 43–106. [53]

Ford, L.R.
   Automorphic Functions, Chelsea Publ. Co., New York, 1951. [36, 197,
   207]

Gavrilov, V.I.
   (a)   Sequential limits of normal meromorphic functions, *Dokl. Akad.
   Nauk SSSR* **138** (1961), 16–17. [182]
   (b)   On the set of angular limiting values of normal meromorphic
   functions, *Dokl. Akad. Nauk SSSR* **141** (1961), 525–526. [182]

Gehring, F.W., Martin, G.J.
   Discrete quasiconformal groups I, *Proc. London Math. Soc.* (3) **55**
   (1987), 331–358. [202]

Goluzin, G.M.
   Geometric Theory of Functions of a Complex Variable, Translations
   of Mathematical Monographs **26**, Amer. Math. Soc., Providence, RI,
   1969. [51, 70, 112, 167]

Grötzsch, H.
   Über das Parallelschlitztheorem der konformen Abbildung schlichter
   Bereiche, *Ber. Verhandl. Kön. Sächs. Gesell. Wiss. Leipzig* **84** (1932),
   15–36. [163]

Hardy, G.H.
   A theorem concerning harmonic functions, *J. London Math. Soc.* **1**
   (1926), 130–131. [191]

Hardy, G.H., Ingham, A.E., Pólya, G.
   Notes on moduli and mean values, *Proc. London Math. Soc.* (2) **27**
   (1928), 401–409. [64]

Hayman, W.K.

   Some remarks on Schottky's theorem, *Proc. Cambridge Phil. Soc.* **43** (1947), 442–454. [58]

   On Nevanlinna's second fundamental theorem and extensions, *Rend. Circ. Mat. Palermo (2)* **2** (1953), 346–392. [123]

   Uniformly normal families, *Lectures on functions of a complex variable*, University of Michigan Press, Ann Arbor, 1955, pp. 199–212. [viii, 82, 88]

   Multivalent Functions, Cambridge University Press, Cambridge, 1958. [164]

   Picard values of meromorphic functions and their derivatives, *Ann. Math.* **70** (1959), 9–42. [132, 136, 142, 145]

   On the limits of moduli of analytic functions, *Ann. Polon. Math.* **12** (1962), 143–150. [64]

   Meromorphic Functions, Oxford University Press, London, 1964. [21, 26, 29, 31, 75, 82, 83, 88, 115, 145, 152, 158]

   Research Problems in Function Theory, Athlone Press of University of London, London, 1967. [113, 142]

   Value distribution of functions regular in the unit disk, *Value Distribution Theory*, Proceedings, Joensuu 1981, pp. 13–43, Lecture Notes in Mathematics, No. 981, Springer-Verlag, Berlin, 1983. [51]

   Subharmonic Functions, Vol. 2, Academic Press, London, 1989. [51, 58, 67]

Helms, L.L.

   Introduction to Potential Theory, Wiley-Interscience, New York, 1969. [183, 185, 193]

Hempel, J.

   The Poincaré metric on the twice punctured plane and the theorems of Landau and Schottky, *J. London Math. Soc. (2)* **20** (1979), 435–445. [94]

   Precise bounds in the theorems of Schottky and Picard, *J. London Math. Soc. (2)* **21** (1980), 279–286. [94]

Hennekemper, G.

   Über Fragen der Werteverteilung von Differentialpolynomen meromorpher Funktionen, Dissertation, Hagen, 1979. [144]

Herglotz, G.

   Über Potenzreihen mit positivem reellen Teil im Einheitskreis, *Ber. Verhandl. Kön. Sächs. Gesell. Wiss. Leipzig* **63** (1911), 501–511. [183]

Hewitt, E., Stromberg, K.

   Real and Complex Analysis, Springer-Verlag, New York 1965. [46]

Hilbert, D.

   Über das Dirichlet Prinzip, *Math. Ann.* **59** (1904), 161–186. [viii]

222   References

Hille, E.
    Analytic Function Theory, Vol. I, II , Ginn and Co., London, 1962.
    [1, 16, 113]

Hiong, K.
    (a)   Sur les fonctions holomorphes dont les déri'vées admettant une
    valeur exceptionnelle, *Ann. École Norm. Sup.* (3) **72** (1955), 165–197.
    [115]
    (b)   Sur les fonctions holomorphes dans le cercle-unité admettant
    un ensemble de valeurs déficientes, *J. Math. Pures Appl.* **34** (1955),
    303–335. [117]
    Sur le cycle de Montel – Miranda dans la théorie des familles nor-
    males, *Sci. Sinica* **7** (1958), 987–1000. [128]
    On the Montel-Miranda cycle in the theory of normal families, *Chi-
    nese Math.* **9** (1967), 390–399. [128]

Hiong, K., Ho, Y.
    Sur les valeurs multiples des fonctions méromorphes et de leurs déri-
    vées, *Sci. Sinica* **10** (1961), 267–285. [58, 150]

Hua, X.
    Normal families of meromorphic functions, preprint (1992). [77, 144]

Hua, X., Chen, H.
    Normal families and the spherical derivatives, preprint (1992). [75,
    133]

Jentzsch, R.
    Untersuchungen zur Theorie der Folgen analytischen Funktionen,
    Acta Math. **41** (1918), 219–251. [44]

Julia, G.
    Mémoire sur l' iteration des fractions rationnelles, *J. Math.
    Pures Appl.* (8) **1** (1918), 47–245. [viii]
    Leçons sur les fontions uniformes à point singulier essential isolé,
    Gauthier-Villars, Paris, 1924. [61, 94]

Khintchine, A.I.
    On sequences of analytic functions, *Recueil Math. Moscou (Mat.
    Sbornik)* **31** (1922–24), 147–151 (Russian). [47]
    Sur les suites de fonctions fonctions analytiques bornées dans leur
    ensemble, *Fund. Math.* **4** (1923), 72–75. [47]

Koebe, P.
    Über die Uniformisierung beliebiger analytischen Kurven, Dritte Mit-
    teilung, *Göttingen Nachr.* (1908), 337–358; [36] Vierte Mitteilung,
    ibid. (1909), 324–361. [185]

Ku, Y.
Sur les familles normales de fonctions méromorphes, *Sci. Sinica* **21** (1978), 431–445. [142]
A criterion for normality of families of meromorphic functions (Chinese), *Sci. Sinica* (special issue) **1** (1979), 267–274. [144]

Kunugui, K.
Sur l'allure d'une fonction analytique uniforme au voisinage d'un point frontière de son domaine de définition, *Japan. J. Math.* **18** (1942), 1–39. [47]

Lai, W.T.
The exact value of Hayman's constant in Landau's theorem, *Sci. Sinica* **22** (1979), 129–134. [94]

Landau, E.
Über eine Verallgemeinerung der Picardschen Satzes, *Sitz. Kön. Preuss Akad. Wiss. Berlin* **38** (1904), 1118–1133. [59]
Zum Koebeschen Verzerrungssatz, *Rend. Circ. Mat. Palermo* **46** (1922), 347–348. [66]
Über die Blochsche Konstante und zwei verwandte Weltkonstanten, *Math. Zeit.* **30** (1929), 608–634. [113]

Langley, J.
On normal families and a result of Drasin, *Proc. Roy. Soc. Edin.* **98**A (1984), 385–393. [137, 150]

Lappan, P.A.
Non–normal sums and products of unbounded normal functions, *Mich. Math. J.* **8** (1961), 187–192. [182]

Larcher, H.
A necessary and sufficient condition for a discrete group of linear fractional transformations to be discontinuous, *Duke Math. J.* **30** (1963), 433–436. [206]

Lattès, S.
Sur l'itération des substitutions rationnelles et les fonctions de Poincaré, *C.R. Acad. Sci. Paris* **166** (1918), 26–28. [176]

Lavrentieff, M.
Sur un problème de M.P. Montel, *C.R. Acad. Sci. Paris* **184** (1927), 1634. [190]

Leau, L.
Les suites de fonctions en général (domaine complexe), Gauthier-Villars, Paris, 1932. [212]

Lebesgue, H.
Sur le problème de Dirichlet, *Circ. Math. Palermo* **24** (1907), 371–402. [189]

Lehner, J.
Discontinuous Groups and Automorphic Functions, Math. Surveys, No. VIII, Amer. Math. Soc., Providence, RI, 1964. Reprinted with corrections 1982. [194]

Lehto, O., Virtanen, K.I.
(a) Boundary behavior and normal meromorphic functions, *Acta Math.* **97** (1957), 47–65. [viii, 178, 181]
(b) On the behaviour of meromorphic functions in the neighbourhood of an isolated singularity, *Ann. Acad. Sci. Fenn. Ser. A.I.* **240** (1957), 9pp. [106]

Li, S.
The normality criterion of a class of the functions, *J. East China Normal Univ.* **2** (1984), 156–158. [137, 151]

Li, X.
The proof of Hayman's conjecture on normal families, *Sci. Sinica* (*Ser. A*) (6) **28** (1985), 596–603. [137]

Lindelöf, E.
Mémoire sur certaines inégalités dans la théorie des fonctions monogènes et sur quelques propriétés nouvelles de ces fonctions dans le voisinage d'un point singulier essentiel, *Acta Soc. Sci. Fenn.* **35** (no. 7) (1909). [62, 63]
Démonstration nouvelle d'un théorème fondamental sur les suites de fonctions monogènes, *Bull. Soc. Math. France* **41** (1913), 171–178. [44]
Sur un principe général de l'analyse et ses applications à la théorie de la représentation conforme, *Acta Soc. Sci. Fenn.* **46** (no. 4) (1915), 1–35. [62]

Lohwater, A.J., Pommerenke, Ch.
On normal meromorphic functions, *Ann. Acad. Sci. Fenn. A.I.* **550** (1973). [101, 181]

Mandelbrojt, S.
Sur les suites de fonctions holomorphes, *J. Math. Pures Appl.* (9) **8** (1929), 173–195. [39]

Mandelbrot, B.
The Fractal Geometry of Nature, Freeman, San Francisco, 1982. [175]

Mañé, R., Sad, P., Sullivan, D.
    On the dynamics of rational maps, *Ann. Sci. École Norm. Sup.* **16** (1983), 193–217. [177]

Martin, R.S.
    Minimal positive harmonic functions, *Trans. Amer. Math. Soc.* **49** (1941), 137–172. [193]

Marty, F.
    Recherches sur la répartition des valeurs d'une fonction méromorphe, *Ann. Fac. Sci. Univ. Toulouse* (3) **23** (1931), 183–261. [75, 211]

Milloux, H.
    Les fonctions méromorphes et leurs dérivées, Hermann et Cie, Paris, 1940. [113, 144]
    Une application de la théorie des familles normales, *Bull. Sci. Math.* (2) **72** (1948), 12–16. [45]

Minda, D.
    Bloch constants, *J. Analyse Math.* **41** (1982), 54–84. [112, 113]
    A heuristic principle for a nonessential isolated singularity, *Proc. Amer. Math. Soc.* (3) **93** (1985), 443–447. [106]
    Another approach to Picard's theorem and a unifying principle in geometric function theory, preprint (1991); to appear in *Current Topics in Analytic Function Theory*, H.M. Srivastava, S. Owa, Editors, World Scientific Publishing Co., Singapore. [108]

Miranda, C.
    Sur un nouveau critère de normalité pour les familles de fonctions holomorphes, *Bull. Soc. Math. France* 63 (1935), 185–196. [101, 125]

Misiurewicz, M.
    On iterates of e^z, *Ergodic Theory Dynamical Systems* **1** (1981), 103–106. [178]

Monna, A.F.
    Dirichlet's Principle, A mathematical comedy of errors and its influence on the development of analysis, Oosthoek, Scheltema and Holkema, Utrecht, 1975. [48]

Montel, P.
 Sur les suites infinies de fonctions, *Ann. École Norm. Sup.* (3) **24**
 (1907), 233–334. [33, 35, 38, 47]
 Leçons sur les series de polynômes à une variable complexe, Gauthier-
 Villars, Paris, 1910. [36, 45]
 Sur l'indétermination d'une fonction uniforme dans les voisinages de
 ses points essentiels, *C.R. Acad. Sci. Paris* **153** (1911), 1455–1456.
 [33]
 Sur les familles de fonctions analytiques qui admettent des valeurs ex-
 ceptionnelles dans un domaine, *Ann. École Norm. Sup.* (3) **29** (1912),
 487–535. [vii, 54, 56, 60, 62]
 Sur les familles normales de fonctions analytiques, *Ann. École Norm.
 Sup.* (3) **33** (1916), 223–302. [74, 79]
 Sur la représentation conforme, *J. Math. Pures Appl.* (7) **3** (1917),
 1–54. [47]
 Sur les familles quasi-normales de fonctions holomorphes, *Mem. A-
 cad. Roy. Belgique* (2) **6** (1922), 1–41. [209]
 Sur les familles quasi-normales de fonctions analytiques, *Bull. Soc.
 Math. France* **52** (1924), 85–114. [209]
 *Leçons sur les familles normales de fonctions analytiques et leurs ap-
 plications*, Gauthier-Villars, Paris, 1927. Reprinted by Chelsea Publ.
 Co., New York, 1974. (viii, 47, 105, 106, 209, 213]
 Sur les séries de fractions rationnelles, *Publ. Math. Univ. Belgrade* **1**
 (1932), 157–169. [46]
 Leçons sur les fonctions univalentes ou multivalentes, Gauthier-
 Villars, Paris, 1933. [65]
 Le rôle des familles normales, *L'Enseignement Math.* **33** (1934), 5–21.
 [84, 125]
 Sur quelques familles de fonctions harmoniques, *Fund. Math.* **25**
 (1935), 388–407. [106, 186, 191]
 Sur les critères de familles normales, *Bull. Sci. Math.* (2) **60** (1936),
 240–246. [211]
 Sur les fonctions localement univalentes ou multivalentes, *Ann. École
 Norm. Sup.* (3) **54** (1937), 39–54. [213]
 Sur les suites de fonctions non bornées dans leur ensemble, *Bull. Soc.
 Math. France* **67** (1939), 42–55. [211]
 Sur le rôle des familles de fonctions dans l'analyse moderne, *Bull.
 Soc. Roy. Sci. Liége* **15** (1946), 262–267. [33]

Mues, E.
 Über ein Problem von Hayman, *Math. Zeit.* **164** (1979), 239–259.
 [108, 136, 142]

Myrberg, L.
Über die Existenz der Greenschen Funktion der Gleichung $\Delta u = c(P)u$ auf Riemannschen Flächen, *Ann. Acad. Sci. Fenn. A.I.* **170** (1954). [194]

Nakai, M.
Existence of positive harmonic functions, *Proc. Amer. Math. Soc.* (2) **17** (1966), 365–367. [194]

Nehari, Z.
Conformal Mapping, McGraw-Hill, New York, 1952. [167]

Nevanlinna, F., Nevanlinna, R.
Über die Eigenschaften analytischer Funktionen in der Umgebung einer singulären Stelle oder Linie, *Acta Soc. Sci. Fenn.* (5) **50** (1922). [47]

Nevanlinna, R.
Zur Theorie der meromorphen Funktionen, *Acta Math.* **46** (1925), 1–99. [21]
Analytic Functions, Springer-Verlag, New York, 1970. [21, 25, 26]

Noshiro, K.
Contributions to the theory of meromorphic functions in the unit circle, *J. Fac. Sci. Hokkaido Univ.* **7** (1938), 149–159. [viii, 178, 179]
Cluster Sets, Springer-Verlag, Berlin, 1960. [181]

Ohtsuka, M.
Generalizations of Montel-Lindelöf's theorem on asymptotic values, *Nagoya Math. J.* **10** (1956), 129–163. [64]

Osgood, W.F.
Note on the functions defined by infinite series whose terms are analytic functions of a complex variable; with corresponding theorems for definite integrals, *Ann. Math.* (2) **3** (1901–02), 25–34. [46]

Oshkin, I.B.
On a test of the normality of families of holomorphic functions, *Uspehi Mat. Nauk* (2) **37** (1982), 221–222 (*Russian Math. Surveys* (2) **37** (1982), 237–238). [133, 151]

Ostrowski, A.
Über die Bedeutung der Jensenschen Formel für einige Fragen der komplexen Funktiontheorie, *Acta Litt. Scient. Univ. Hung.* **1** (1922–23), 80–87. [47]
Über Folgen analytischer Funktionen und einige Verschärfungen des Picardschen Satzes, *Math. Zeit.* **24** (1926), 215–258. [71, 74, 212]

228   References

Palka B.P.
   An Introduction to Complex Function Theory, Springer-Verlag, New York, 1991. [1, 95, 154, 167]

Pang, X.
   Bloch's principle and normal criterion, *Sci. Sinica* (7) **32** (1989), 782–791. [103, 104, 107, 137, 143]
   On normal criterion of meromorphic functions, *Sci. Sinica* (5) **33** (1990), 521–527. [104, 107, 108, 143]

Peitgen, H.O., Richter, P.H.
   The Beauty of Fractals, Springer-Verlag, Berlin, 1986. [167]

Pfluger, A.
   Lectures on Conformal Mapping, Lecture Notes, Dept. of Mathematics, Indiana University, 1969. [70]

Picard, É.
   (a)   Sur une propriété des fonctions entières, *C.R. Acad. Sci. Paris* **88** (1879), 1024–1027. [56]
   (b)   Sur les fonctions analytiques uniformes dans le voisinage d'un point singulier essentiel, *C.R. Acad. Sci. Paris* **89** (1879), 745–747. [60]

Piranian, G., Thron, W.J.
   Convergence properties of sequences of linear fractional transformations, *Mich. Math. J.* **4** (1957), 129–135. [201]

Pólya, G.
   Über die Nullstellen sukzessiver Derivierten, *Math. Zeit.* **12** (1922), 36–60. [152]

Pommerenke, Ch.
   On Bloch functions, *J. London Math. Soc.* (2) **2** (1970), 689–695. [181]
   Univalent Functions, Vandenhoeck and Ruprecht, Göttingen, 1975. [40, 68]

Porter, M.B.
   Concerning series of analytic functions, *Ann. Math.* (2) **6** (1904–05), 190–192. [44]

Radó, T.
   Über die Fundamentalabbildung schlichter Gebiete, *Acta Szeged* **1** (1922–23), 240–251. [51]
   Subharmonic Functions, Springer-Verlag, Berlin, 1937. Reprint 1971. [39]

Riesz, F.
Sur les suites de fonctions analytiques, *Acta Litt. Scient. Univ. Hung.*
1 (1922-1923), 88–97. [47]

Robinson, A.
Metamathematical problems, *J. Symbolic Logic* **38** (1973), 500–516.
[vii, 101, 103]

Ros, A.
The Gauss map of minimal surfaces, preprint (1992). [104]

Royden, H.
A criterion for the normality of a family of meromorphic functions,
Ann. Acad. Sci. Fenn. A.I. **10** (1985), 499–500. [76]

Rubel, L.
Four counterexamples to Bloch's principle, *Proc. Amer. Math. Soc.*
98 (1986), 257–260. [107]

Rudin, W.
Real and Complex Analysis, 2nd ed., McGraw-Hill, New York, 1974.
[47]

Sansone, G., Gerretsen, J.
Lectures on the Theory of Functions of a Complex Variable, Vol. II,
Wolters-Noordhoff, Groningen, 1969. [113]

Saxer, W.
Ueber quasi-normale Funktionsscharen und eine Verschärfung des Pi-
cardschen Satzes, *Math. Ann.* **99** (1928), 708–737. [213]
Ueber konvergente Folgen meromorpher Funktionen, *Comm. Math.
Helv.* **2** (1930), 18–34. [211]

Schottky, F.
Über den Picardschen Satz und die Borelschen Ungleichungen, *Sitz.
Kon. Preuss. Akad. Wiss. Berlin* **42** (1904), 1244–1262. [57, 60]

Schwick, W.
Normality criteria for families of meromorphic functions, *J. Analyse
Math.* **52** (1989), 241–289. [87, 123, 133, 143, 151, 152, 158]
An estimation of the proximity function of the logarithmic derivative
for families of meromorphic functions, *Complex Variables Theory Ap-
pl.* (3) **15** (1990), 149–154. [122]

Segal, S.
Nine Introductions in Complex Analysis, North-Holland, Amsterdam,
1981. [16, 94]

Selecta :
    Cinquantenaire Scientifique de M. Paul Montel, Gauthier-Villars,
    Paris, 1947. [209]

Shimizu, T.
    On the theory of meromorphic functions, *Japan. J. Math.* **6** (1929),
    119–171. [26]

Stieltjes, T.J.
    Recherches sur les fractions continues, *Ann. Fac. Sci. Toulouse* **8**
    (1894), 1–22. [45]

Stroyan, K.D., Luxemburg, W.A.J.
    Introduction to the Theory of Infinitesimals, Academic Press, New
    York, 1976. [104, 106]

Tsuji, M.
    Potential Theory in Modern Function Theory, Maruzen Co. Ltd.,
    Tokyo, 1959. [189]

Valiron, G.
    Sur un théorème de M M. Koebe et Landau, *Bull. Sci. Math.* (2) **51**
    (1927), 34–42. [64]
    Familles normales et quasi-normales de fonctions méromorphes,
    Gauthier-Villars, Paris, 1929. [82, 105, 209]

Vitali, G.
    Sopra le serie di funzioni analitiche, *Rend. della R. Inst. Lombardo di
    Sci. Lett.* (2) **36** (1903), 772–774. [44]
    Sopra le serie di funzioni analitiche, *Ann. Mat. Pura Appl.* (3) **10**
    (1904), 65–82. [44]

Yang, L.
    Sur les valeurs quasi-exceptionnelles des fonctions holomorphes, *Sci.
    Sinica* **13** (1964), 879–885. [66]
    Meromorphic functions and their derivatives, *J. London Math. Soc.*
    (2) **25** (1982), 288–296. [144]
    Normal families and differential polynomials, *Sci. Sinica, Ser. A* **26**
    (1983), 673–686. [159]
    A general criterion for normality, *Acta Math. Sinica (New Series)* **1**
    (1985), 181–193. [151]
    (a) Normal families and fix-points of meromorphic functions, *Indiana
    Univ. Math. J.* **35** (1986),179–191. [150]
    (b) Normality for families of meromorphic functions, *Sci. Sinica,
    Ser. A* **29** (1986), 1263–1274. [150]

Yang, L., Chang, K.
    Recherches sur la normalité des familles de fonctions analytiques à

des valeurs multiples,

I. Un nouveau critère et quelques applications, *Sci. Sinica* **14** (1965), 1258–1271, [58, 65, 66, 132]

II. Géneralizations, *Sci. Sinica* **15** (1966), 433–453. [105, 150]

Ye, Y.

A new normal criterion and its application, *Chinese Ann. Math. Ser. A* (Supplement) **12** (1991), 44–49. [137]

Yosida, K.

On a class of meromorphic functions, *Proc. Phys. Math. Soc. Japan* **16** (1934), 227–235. [178]

Zalcman, L.

A heuristic principle in complex function theory, *Amer. Math. Monthly* **82** (1975), 813–817. [vii, 101, 103]

Modern perspectives on classical function theory, *Rocky Mountain J. Math.* (1) **12** (1982), 75–92. [103]

Index